T0275674

THE ELECTRICAL PROPERTIES OF DISORDERED METALS

Cambridge Solid State Science Series

EDITORS:
Professor I. M. Ward, FRS
Department of Physics, University of Leeds
Professor D. R. Clarke
Department of Materials Science and Engineering
University of California, Santa Barbara

The theory of how solid metals conduct electricity has, until recently, been confined to crystalline metals in which the constituent atoms form regular arrays. The discovery of how to make solid amorphous metallic alloys (often called metallic glasses), in which the atoms are no longer ordered, led to an explosion of measurements on these new materials. A whole range of new and unexpected behaviour was found, particularly at low temperatures and in a magnetic field. At the same time theories to explain the electrical properties of disordered metals began to emerge.

To understand this new behaviour, conventional Boltzmann theory, which assumes that the free path of the conduction electrons is long and only occasionally interrupted by scattering, has to be extended and modified when the mean free path becomes comparable with the wavelength of the electrons and with the distance between neighbouring atoms.

The theory is explained in physical terms and the results are compared to experimental results on metallic glasses.

The book is designed to be self-contained and to appeal to non-specialist physicists, metallurgists and chemists with an interest in disordered metals or to students beginning to study these materials.

THE ELECTRICAL PROPERTIES
OF DISORDERED METALS

J. S. DUGDALE

Emeritus Professor of Physics
University of Leeds, UK

CAMBRIDGE
UNIVERSITY PRESS

CAMBRIDGE UNIVERSITY PRESS
Cambridge, New York, Melbourne, Madrid, Cape Town, Singapore, São Paulo

Cambridge University Press
The Edinburgh Building, Cambridge CB2 2RU, UK

Published in the United States of America by Cambridge University Press, New York

www.cambridge.org
Information on this title: www.cambridge.org/9780521268820

First published 1995
This digitally printed first paperback version 2005

A catalogue record for this publication is available from the British Library

Library of Congress Cataloguing in Publication data
Dugdale, J. S. (John Sydney).
The electrical properties of disordered metals / J.S. Dugdale.
p. cm. – (Cambridge solid state science series)
Includes bibliographical references and index.
ISBN 0-521-26882-6
1. Free electron theory of metals. 2. Metals – Electric
properties. 3. Metallic glasses – Electric properties. 4. Order
–disorder models. I. Title. II. Series.
QC176.8.E4D78 1995
530.4′13 – dc20 95–1879 CIP

ISBN-13 978-0-521-26882-0 hardback
ISBN-10 0-521-26882-6 hardback

ISBN-13 978-0-521-01751-0 paperback
ISBN-10 0-521-01751-3 paperback

KW

To my wife

Contents

Preface

The purpose of this book is to explain in physical terms the many striking electrical properties of disordered metals or alloys, in particular metallic glasses. The main theme is that one central idea can explain many of the otherwise puzzling behaviour of these metals, particularly at low temperatures and in a magnetic field. That idea is that electrons in such metals do not travel ballistically between comparatively rare scattering events but diffuse through the metal. These new effects are not large but they are so universal in high-resistivity metals, so diverse and qualitatively so different from anything to be expected in metals where the electrons have a long mean free path, that they cry out for an explanation.

The book is not a critical research review; the motivation is mainly to explain. In interpreting theory there are always the dangers of overinterpretation, misinterpretation and failure to interpret and I do not expect to have escaped these completely. Nonetheless, our new understanding of disordered metals and alloys constitutes a substantial addition to conventional Boltzmann theory and deserves to be more widely known and appreciated.

The book is aimed at those who know little of the subject such as students starting work in this field or those outside the field who wish to know of developments in it. There is no attempt at rigorous derivations; the aim is to present the physics as clearly as possible so that readers can think about the subject for themselves and be able to apply their thinking in new contexts.

For those whose knowledge of electron transport properties is limited to what they learned in undergraduate courses I outline briefly the main points of conventional theory in the first part of the book. This is not meant to be an exhaustive treatment but a reminder of ideas already encountered and here put in the context of what is to follow.

I am greatly indebted to many friends and colleagues for discussions, reading the manuscript or parts of it and help in understanding the subject. I cannot attempt to mention them all but I am particularly grateful to the following: Denis Greig (who introduced me to the subject), Bryan Gallagher, Bryan Hickey, Mark Howson, Jim Morgan and Davor Pavuna, with all of whom I had the pleasure of working on aspects of this subject. I found the thesis of Dr A. Sahnoune most clear and helpful. I am indebted to Dr Moshe Kaveh for reading the manuscript and to Nathan Wiser for helpful discussions. I must also record my deep gratitude to the editor, Ian Ward, for much help and encouragement; to Dr B. M. Watts, who as copy editor did so much to improve the manuscript; and to Mrs Mary Edmundson for her tireless help in preparing drafts and organising the related correspondence.

To all these and to many others, my sincere thanks.

J. S. Dugdale
November 1994

1

Context and content

Introduction

1.1 Ordered crystalline metals

Our understanding of the electrical conductivity of metals began almost a century ago with the work of Drude and Lorentz, soon after the discovery of the electron. They considered that the free electrons in the metal carried the electric current and treated them as a classical gas, using methods developed in the kinetic theory of gases.

A major difficulty of this treatment was that the heat capacity of these electrons did not appear in the experimental measurements. This difficulty was not cleared up until, in 1926, Pauli applied Fermi–Dirac statistics to the electron gas; this idea, developed by Sommerfeld and his associates, helped to resolve many problems of the classical treatment. The work of Bloch in 1928 showed how a fully quantal treatment of electron propagation in an ordered structure could explain convincingly many features of the temperature dependence of electrical resistance in metals. In particular it showed that a pure, crystalline metal at absolute zero should show negligible resistance.

From these beginnings followed the ideas of the Fermi surface, band gaps, Brillouin zones, umklapp processes and the development of scattering theories: the scattering of electrons by phonons, impurities, defects and so on. By the time of the Second World War, calculations of the resistivity of the alkali metals showed that the theory was moving from qualitative to quantitative success.

In 1950, the recognition that the de Haas–van Alphen effect provided a measure of the extremal cross-section of the Fermi surface normal to the applied magnetic field made possible a big advance in the experimental study of Fermi surfaces. This was matched by a corresponding development of their theoretical calculation.

The isotope effect in superconductivity was discovered about this time and provided the clue that the electron–phonon interaction was implicated in the phenomenon. In 1957 the BCS theory of superconductivity, giving an explanation of many of its aspects, was published by Bardeen, Cooper and Schrieffer. This was a problem that had for a long time resisted a satisfying theoretical understanding[1].

Experimental improvements made possible the measurement of electron velocities over the Fermi surface and also the electron lifetimes under different scattering mechanisms. By putting together all this information within the theoretical framework that had been developed it was possible, by the 1970s, to explain the electrical conductivity of simple metals in considerable detail. But this was not universally accepted; there was a protracted argument as to whether the ground state of, for example, potassium, involved charge density waves. Most people involved finally concluded that the answer was no.

Transition metals, having the added complication of an incompletely filled d-band, had not received such detailed study either by experiment or theory. Nonetheless, it was felt that the same experimental methods and theory that had been so successful for the simple metals could lead to a comparable understanding of the pure transition metals if the effort were forthcoming. In short, pure crystalline metals were thought to be understood[2].

1.2 Disordered metals

The simplest examples of disordered metals are the liquid alkali metals and though their electrical properties had been measured, the understanding of their conductivity had not made much progress until the Ziman theory of 1961. The problem in a liquid metal is twofold: How do you describe the structure of the ions in the liquid and how do you calculate the scattering of conduction electrons from them? Ziman exploited the use of the structure factor to answer both questions. The diffraction of neutrons and X-rays yields the structure factor of liquid metals; this gives directly the probability of scattering of the incident waves as a function of scattering angle. These diffraction experiments

[1,2] See *Notes*, commencing on p.223.

are so designed that as far as possible only one scattering event takes place within the target material and so these results can also be used to describe the scattering of plane-wave electrons. Moreover, although the structure of liquids cannot be described theoretically in a fully satisfactory way, there are approximate methods that allow one to calculate it so that experimental or theoretical structure factors of the metals provide a fairly direct way of calculating the electrical resistivity.

The Hall coefficient of most simple metals is free-electron-like. Therefore the dynamics of the electrons in the liquid are straightforward and all that is required in addition to the structure factor is the form factor or scattering cross-section of the appropriate ions. This is usually deduced from the pseudopotential of the ion, which in turn can be calculated or found semi-empirically from suitable measurements on the corresponding crystal.

The Ziman theory and its development have been able to account for the magnitude and temperature coefficient of the resistivity of a number of simple metals as well as the thermoelectric power (thermopower, for short) and pressure coefficient of resistivity. It can also explain the systematic differences between the temperature coefficients of resistance of monovalent and polyvalent simple metals. It has, however, been less successful with liquid transition metals.

A different class of disordered metals is provided by alloys. Let us consider for simplicity an alloy of two components, which we can call A and B. A crystalline alloy can be formed either in an ordered structure with appropriate numbers of A and B ions corresponding to its molecular composition (e.g. A_2B_3) or in a disordered but still crystalline structure, in which the two (or more) components are distributed randomly on the lattice sites of the crystal. The latter are sometimes referred to as random solid solutions and it is these that concern us here; the ordered alloys can be treated in much the same way as single component crystals. Calculations on, for example, the silver–palladium series, which form continuous random solid solutions right across the composition range, have indicated that notions such as the Fermi surface and k-vector derived from ordered structures can still be useful in such materials, and experiments with angle-resolved photo-emission have tended to support this view. Calculations of electron transport properties by means of theories derived originally for ordered metals have been reasonably successful, at least in broad outline.

1.3 Beyond Boltzmann theory

All these successes of our theoretical understanding, with the exception of
the BCS theory of superconductivity, have been achieved within the
compass of what is usually referred to as Boltzmann theory. Essentially
this means that the mean free path of the conduction electrons is assumed
to be long compared to the wavelength of the electrons at the Fermi level.
This in turn implies that one scattering event is independent of another
and any interference between the scattered wave and the wave before
scattering can be ignored. If, however, the mean free path is very short
as it is in highly disordered metals or alloys, this interference can become
important. This book is largely concerned with the consequences of this
new situation.

1.4 Metallic glasses

The discovery of how to make solid amorphous alloys by rapid quench-
ing from the melt led to an explosion of measurements of the electronic
properties of these new materials. These alloys are generally referred to as
metallic glasses because like window glass they have a structure that
resembles that of a liquid when the constituent atoms are frozen in
their instantaneous positions. As a consequence of the high disorder
the conduction electrons have a very short mean free path and thus
their behaviour does not correspond with Boltzmann theory. The measurements on metallic glasses coincided with the extension of the theory
to take account of these interference effects and so, at least for bulk
materials, metallic glasses became the testing ground for the new theories.
In what follows, therefore, I shall try to explain what the theories have to
say and consider how far they account for the experimental measurements on metallic glasses.

As we have seen, liquid metals have been much studied as examples of
highly disordered metals but because they are liquids they suffer from a
number of severe disadvantages as experimental subjects. They do not
exist at low temperatures; indeed their range of stability (between freezing
point T_F and boiling point T_B) is very limited when expressed as a ratio
T_B/T_F. They are subject to convection currents when heated and can be
corrosive and difficult to handle and contain.

Metallic glasses, although unstable at higher temperatures because they
revert to the crystalline phase, can be studied at low temperatures where
many of the most interesting phenomena reveal themselves. As it turns

out many alloys, albeit in restricted ranges of composition, are fairly easy to make in ribbon form, which is very suitable for measurement of their electrical properties such as conductivity, Hall coefficient and thermopower.

In order to understand the methods of making metallic glasses and the limitations that their stability in terms of composition and temperature range imposes, I outline some of the main features of the production and structure of these glasses in the next chapter. Thereafter the important ideas of the Boltzmann theory of electrical conduction are explained, culminating in Chapter 5 in their application in the Ziman theory to the electrical properties of simple liquid metals.

In Chapter 6 the specifically low-temperature behaviour of metals including the electron–phonon interaction is examined since these features are not involved in the Ziman theory; in Chapter 7 the notion of quasi-particles and interactions between electrons are discussed and in Chapter 8 the properties of transition metals are outlined. Then as a final consideration of Boltzmann theory the Hall effect and magnetoresistance are considered because these properties are very important to our understanding of high-resistivity materials.

Chapters 11 to 14 then concentrate on how recent theories have gone beyond Boltzmann theory in their attempts to explain a wide range of unusual low-temperature behaviour. The theories are here applied specifically to the resistivity and magnetoresistance of metallic glasses. Before finally attempting a quantitative comparison of experiment and theory, Chapter 15 is concerned with the thermopower of metallic glasses, a property that gives valuable information about the behaviour of electrons but is not so directly responsive to the new interference and interaction effects. Chapter 16 provides a comparison between theory and experiment in a selection of metallic glasses; it attempts to show how far the new theories can account naturally and convincingly for electron conduction in highly disordered metallic conductors.

I have made no attempt to deal with strongly magnetic metals, whose electrical properties, even in the crystalline state, are still not fully understood; nor have I discussed the two-level systems found in glasses, since they appear to make no significant contribution to the properties of metallic glasses in the temperature range discussed here.

2

Production and structure of metallic glasses

2.1 What are metallic glasses?

The word 'glass' as we normally use it refers to window glass. As we all know, this is a brittle, transparent material with vanishingly small electrical conductivity. It is in fact a material in which the constituent molecules are arranged in a disordered fashion as in a liquid but not moving around; that is to say, each molecule keeps its same neighbours and the glass behaves like a solid. Most of the solids that physicists have hitherto dealt with are crystalline i.e. their atoms or molecules are arranged in strictly ordered arrays. This is the essential difference between a so-called 'glass' and a crystal: *a glass has no long-range order*. Although the word 'glass' was originally used to designate only window glass it has now taken on this generalised meaning of what we may call an amorphous solid.

Electrically insulating glasses have been studied for a long time and it was generally thought that in order to form a glass by cooling a liquid it was necessary to have a material composed of fairly complicated molecules so that, on cooling through the temperature range at which crystallisation would be expected to occur, the molecules would have difficulty in getting into their proper places and could be, as it were, frozen in a disordered pattern at lower temperatures without the thermal energy necessary to get into their ordered positions. This general picture is correct and helpful although the expectations based on it have proved in some respects wrong. It was thought that because metals and alloys are usually of simple atoms, it would be impossible to form a glass from such constituents. It therefore came as a considerable surprise when, in 1959, Pol Duwez and his co-workers at the California Institute of Technology showed that an alloy of gold and silicon could be made to form a glass. The secret in part was to increase the rate of

6

cooling from the melt to such an extent that even with such simple constituents, their atoms did not have time to take up their ordered positions before diffusive motion became impossible through lack of thermal energy.

Since then many alloys have been made to form glasses although, as we shall see, there are limits to the combinations of metals and to the ranges of concentrations for which glass formation is readily possible. These glassy alloys are typically characterised by metallic properties: they conduct electricity comparatively well, they have a shiny appearance and they are ductile i.e. not brittle. In fact they are, as we shall see, truly metallic and this explains why they are called 'metallic glasses'.

2.2 Properties of glasses in general

Before we discuss the question of how to produce metallic glasses, let us first look at some of the properties of glasses in general. Figure 2.1 shows how the volume (per unit mass, say) of a glass-forming material changes with temperature. Consider first what happen when the material is cooled very slowly and crystallises; we assume for simplicity that we are dealing with a system of a single component since this does not alter the essential points. The volume changes abruptly at the freezing point T_F (in almost all systems it contracts as shown in the figure). Thereafter, as it cools further, the volume slowly diminishes as indicated by the dashed line marked A in Figure 2.1. If however we cool the material in such a way that it forms a glass there is no discontinuity at the freezing point and the volume diminishes smoothly through this region as shown by the line marked B in the figure. In this case, however, the path followed and the ultimate volume reached depend on the cooling rate as indicated by the two lines B and C which correspond to different cooling rates, C being the faster.

At sufficiently low temperatures, it is possible to make changes to the state of the glassy material in a perfectly reversible manner by, for example, changes in pressure or temperature. If however the temperature is too high, spontaneous irreversible changes may occur. This is because there is always a tendency for the glass to move towards the truly stable thermodynamic state which is that of the crystal. For this reason the glassy state is an unstable state although it is often referred to as metastable. The term 'metastable' is however better reserved for the supercooled liquid which is a liquid cooled carefully below its freez-

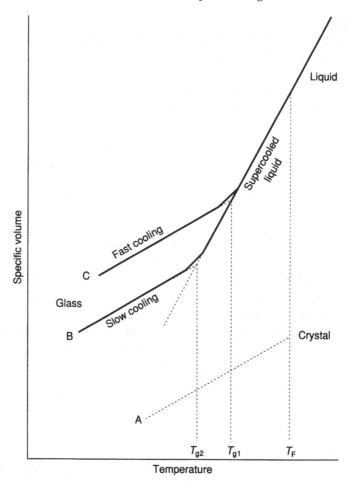

Fig. 2.1 Specific volume of a glass as a function of temperature, showing also the crystal and supercooled liquid. The final volume of the glass depends on the cooling rate.

ing point so as to retain its full liquid properties. Such a liquid can undergo perfectly reversible changes provided that there are no nucleation sites which would trigger crystallisation. When this does occur it is sudden and complete. One may think of the supercooled liquid as separated by a large free energy barrier from the crystalline state. The glassy state, on the other hand, is separated from the crystalline state by a succession of quite small potential barriers each leading to further unstable states which lead ultimately to that of the crystal. The supercooled liquid is as it were at the bottom of a substantial valley whereas the glass is perched rather precariously on the side of a hill at whose base is the crystalline state.

In Figure 2.1 the supercooled liquid is shown as existing below the freezing point and the line that shows the volume of this liquid is continued to low temperatures even though it might in practice be very difficult or even impossible to achieve this metastable state at such temperatures. This line is, however, a useful extrapolation. The temperature at which the line referring to the glass departs from the equilibrium line of the supercooled liquid is called the glass transition temperature T_g. As we see, different cooling rates yield different transition temperatures and in metallic glasses these can differ by as much as 100 K with the corresponding densities differing by up to 0.5%.

If the glass is allowed to warm up from low temperatures, it begins to recrystallise at some temperature which depends on how it was cooled and on how quickly is it warmed. It is therefore clear that the properties of glassy materials depend on their past history. Fortunately the electrical properties in which we are interested are reasonably reproducible if the specimens are made with standard techniques and are not allowed to become too warm. In practice this means confining experiments to systems that are stable at temperatures up to at least room temperature. Even so it is sometimes necessary to store the specimens at low temperatures in, for example, liquid nitrogen when they are not being used for experiments. In general, however, if suitable fairly straightforward precautions are taken, the reproducibility of the measurements on these metallic glasses is quite adequate to produce a clear picture of their electrical properties.

2.3 How are metallic glasses produced?

As we know from experience most liquids tend to crystallise on cooling and as we have already seen, a glass will be formed from the liquid only if we can cool it through the region of the freezing point sufficiently fast. The key to producing metallic glasses is thus to be able to cool them rapidly from the melt. As we shall see, there are other techniques that can be used but we shall first consider those based on rapid cooling because they have been most commonly used and with great success.

2.3.1 Melt spinning

In the melt spinning process a jet of molten alloy is squirted onto a rapidly rotating roller which absorbs the heat and so continuously cools the liquid as it strikes the moving surface. A typical apparatus is illustrated in Figure 2.2. This shows the quartz crucible which has a jet at its bottom end. The alloy is melted by a radiofrequency heating coil around the crucible. The roller is usually made of copper because of its good thermal conductivity and when it is spinning at high speed the alloy is driven onto it through the jet by increasing the pressure of gas (helium or argon, say) above the liquid. In this way a ribbon of metallic glass about 50 μm thick flies off the roller. For a typical charge of order 10 g some 20 or so metres of ribbon are produced in a tenth of a second. It can be quite spectacular. Even more so if it goes wrong! Cooling rates achieved in this way can be as high as a million degrees a second. The ribbons are necessarily thin; the thickness cannot be increased beyond about 100 μm for most alloys as it would be impossible to achieve high

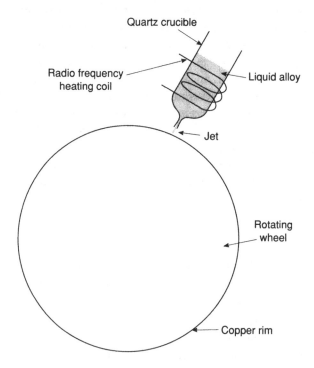

Fig. 2.2 Melt spinning apparatus, showing the copper-rimmed wheel, the molten alloy in the crucible and RF heating coil.

enough cooling rates to form a glass. On the other hand, the width can be increased and in industrial processes sheets up to about 30 cm have been produced.

There are variants on this process with, for example, the use of two rollers in order to cool both surfaces of the ribbon. Another possibility is to enclose the whole apparatus in a container which can be evacuated or filled with a suitable inert gas. This not only protects the alloy from exposure to the atmosphere but can also improve the uniformity of the ribbon. This is because the flow of the liquid alloy into the puddle that forms on the wheel and the emergence of the solid ribbon from it are affected by the viscosity and thermal conductivity of the gas around it.

2.3.2 Atomic deposition techniques

Atomic deposition techniques include several which produce thin films of material, for example, electrolytic and chemical deposition, vacuum deposition on a cold substrate and radiofrequency sputtering. This last has been quite widely used because it can, at least in principle, extend the range of alloys that can be made to form glasses. It does, however, suffer from the disadvantage that it produces rather thin films on a substrate and this has tended to concentrate attention on the thin film aspect of the specimens rather than their bulk character. The technique does, however, have its place in the array of valuable methods for making metallic glasses.

2.4 Alloys that can form stable glasses

It is found by experience that only alloys in a limited range of compositions can be quenched to form glasses that are stable at room temperature. For our purposes it is convenient to divide these into three main classes as follows:

1 Metal–metalloid alloys consisting typically of about 80% of a transition metal such as Fe, Ni, Co and around 20% of one or more metalloid elements: B, C, N, Si, P. Examples are $Fe_{80}B_{20}$ or $Fe_{40}Ni_{40}P_{14}B_6$ where subscripts refer to atomic percentages.

2 A late transition metal combined with an early transition metal, such as $Cu_{50}Ti_{50}$ or $Ni_{60}Nb_{40}$. Here the range of compositions may be rather wider than in class 1. Moreover a rare earth may be substituted for a

transition metal, for example, Gd–Co. This general group of alloys has proved of great interest and value in studying electron transport properties.

3 Alloys of simple i.e. non-transition metals such as $Mg_{70}Zn_{30}$ or $Ca_{30}Al_{70}$. These alloys have a special importance because they are rather simpler to treat theoretically than alloys of transition metals with incomplete d-shells. They will, however, form glasses in only a rather restricted range of compositions.

There are certain features of the classes listed above that can perhaps be understood in fairly simple terms. First of all it is clear that the pure metals do not readily form glasses and possibly not at all. In other words we have to have at least two components. In the alloys of class 1 we find that the compositions that are favourable for glass formation are primarily those close to the eutectic compositions. The eutectic composition has the lowest melting point of all the alloys with compositions near it. The difference between the melting temperature and the glass transition temperature is thus comparatively small so that the system can be cooled very quickly into the glassy state. Moreover a eutectic alloy is an intimate mixture of two distinct crystalline phases. Here I emphasise the word 'mixture' because it shows that the molten alloy refuses to form an ordered compound or a disordered solid solution even when cooled slowly with plenty of time to take up its equilibrium configuration. It is therefore the more willing to form a glass when cooled quickly. Indeed a feature common to all the alloys that form glasses is that they have components and compositions which do not readily form either solid crystalline compounds or crystalline solid solutions. In order to crystallise they would have to form a *mixture* of crystals of very different composition. It seems that these conditions favour glass formation.

An important consequence of this feature of glass formation is that when a glassy alloy crystallises or devitrifies, as it is called, it does not yield a uniform crystal of well-defined character and composition but rather a mixture of crystals of different compositions, probably in the form of a brittle solid or powder. This makes comparison of the glass with its corresponding crystal, a comparison that would often be very revealing, difficult or impossible.

To sum up this picture of glass formation we can say that, for a given cooling rate, a metallic glass tends to form if the forces between the constituent atoms and the relative concentration of the components inhibit crystallisation long enough for the alloy to reach a temperature at which rearrangement of the atoms has become virtually impossible through lack of thermal energy. The glass then is quite similar in structure to the liquid; it is as if the liquid had been suddenly frozen in time.

2.5 Structure of metallic glasses

The X-ray diffraction patterns of crystalline solids show sharp peaks corresponding to Bragg reflections from parallel planes of atoms in the crystal. By contrast metallic glasses show only broad diffuse peaks as indicated in Figure 2.3. This brings out the important structural difference between the crystalline and the glassy or, more generally, the amorphous state. The crystalline state shows long-range order, the amorphous or glassy state does not.

Figure 2.3 shows the diffraction pattern of a metallic glass. Such a pattern, whether derived from X-rays or neutrons, is obtained by allowing a parallel stream of particles characterised by a wave vector $\mathbf{k}$ to fall on the amorphous sample and observing the intensity of the diffracted beam of wave vector $\mathbf{k}'$ as a function of the angle θ between $\mathbf{k}$ and $\mathbf{k}'$. We will confine our attention to elastic scattering which implies that the energies of the particles are the same before and after scattering so that

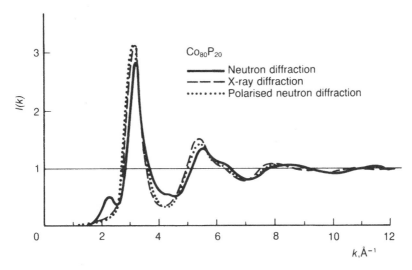

Fig. 2.3 Total scattered intensity of neutrons, polarised neutrons and X-rays diffracted from $Co_{80}P_{20}$. (After Sadoc and Dixmier 1976.)

k and **k'** have the same magnitude. It is often convenient to use as our variable not θ but the scattering vector **K** defined as:

$$\mathbf{K} = \mathbf{k}' - \mathbf{k} \tag{2.1}$$

Figure 2.4 shows this equation in diagrammatic form and the relationship between θ and the magnitude of K:

$$K = 2k_0 \sin(\theta/2) \tag{2.2}$$

where $k_0 = k = k'$.

We can think of the physics of structure determination in the following way. The incident wave **k** is modulated by the scattering field inside the material and if we concentrate on a particular Fourier component of this field, of wave vector **q**, say, this modulation produces a scattered wave of wave vector $\mathbf{k}' = \mathbf{k} + \mathbf{q}$. Rearranged, this gives $\mathbf{q} = \mathbf{k}' - \mathbf{k} = \mathbf{K}$. Thus we see that the scattering of the incident waves from **k** to **k'** explores the structure of the material under study by looking in turn at the strengths of the different Fourier components as **k'** and hence **K** changes. If there is a prominent periodicity **q** in the material this will produce a strong response in the outgoing beam when $\mathbf{K} = \mathbf{q}$. In all this we are assuming that the incoming particle (X-ray or neutron or whatever) undergoes only one scattering event during its passage through the material and indeed the experiments are arranged so that this condition is satisfied. Under

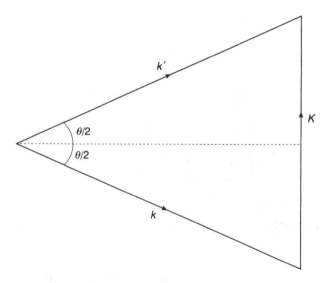

Fig. 2.4 Scattering vector and scattering angle.

these conditions we see that the result of this kind of diffraction experiment is to produce a Fourier analysis of the scattering field inside the material.

2.5.1 The structure factor

In calculating the response of a material to the incident waves we must remember that interference effects lie at the heart of the diffraction pattern so that we must first evaluate the scattering amplitude, which takes account of the phase of the combining waves, before evaluating the square modulus of this, which gives the diffracted intensity. It is the intensity that is measured in a typical experiment.

The total scattering amplitude depends not only on the structure of the material but also on the scattering power of the constituent atoms. Indeed after taking the square modulus of the amplitude to obtain the intensity, it turns out that the intensity of the scattered radiation in any given direction is given by the product of two factors, one of which, the structure factor, depends only on the structure, i.e. only on the positions of the scattering centres, and the other, called the form factor, which depends only on the scattering strength of the individual scatterers. In general we can assume that the form factor is known independently from measurements of the scattering cross-sections or scattering lengths of the individual atoms or nuclei. From the diffraction pattern of the glass we can therefore find the structure factor of the glass, which gives us the Fourier transform of the scattering field.

The structure factor, or interference function, of a metallic glass is a very important feature of the glass so that we need to have some familiarity with it. As I have already emphasised the structure factor of a crystal vanishes except at certain angles where sharp lines occur; these are the Bragg reflections. In a gas, by contrast, the structure factor is everywhere unity (at least insofar as it is an ideal gas of point particles) since the atoms are randomly distributed in space at any instant and so can give rise to no coherent interference effects. The diffraction pattern of a gas is just the sum of the diffracted intensities of the individual atoms.

The structure factor of glasses or liquids (Figure 2.5) is characterised by a large broad peak followed by a few heavily damped oscillations before tending towards unity as K increases. How does this come about? Although there is no long-range order in the amorphous phase there is considerable short-range order imposed by the hard cores of the ions which essentially prevent their overlap. The glass or liquid is like a

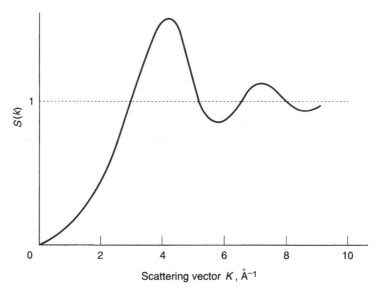

Fig. 2.5 Structure factor of metallic glass (schematic).

fairly close-packed collection of almost hard spheres. There is thus a ring
of first neighbours around any given ion at an average distance of, say,
d_1, a less well-defined ring of second neighbours at d_2 and so on. These
give rise to prominent periodicities and so the first ring produces a peak
in the diffracted intensity whose centre is at about $K = 2\pi/d_1$ and whose
breadth arises from variations in the separation of neighbouring ions; it is
followed by further peaks at the same periodicity but smaller amplitude.
The second ring contributes similar peaks of period $2\pi/d_2$ but more
heavily damped. At small values of K, the structure factor is very small
because at these long wavelengths the material appears homogeneous
with no significant periodicities and only small fluctuations. Thus the
composite pattern starts from the origin at small values, rises to a
broad peak with contributions from the innermost rings and then
shows smaller peaks. At large values of K the probing wavelength has
become so short that the disparities in inter-neighbour distance quickly
destroy any favourable phase coherence. Thus the structure factor tends
to unity as in the gas.

  As we saw earlier the scattering amplitude of the diffracted beam is the
Fourier transform of the scattering field so that the diffraction pattern
picks out the important spatial periodicities of the structure.
Unfortunately as I have already stressed, we measure not amplitudes

but intensities. In consequence we cannot derive the complete structure from the diffraction pattern but only the pair distribution function which in isotropic systems is the radial distribution function (see below).

What I have said so far applies if one type of atom only is present. In an alloy however there are partial structure factors which depend on the type of atom under consideration. For example in a binary alloy AB there is a partial structure factor S_{AA} that depends on the position of A-atoms relative to a central A-atom, a corresponding S_{BB}, and a cross term S_{AB} that depends on the position of B-atoms with respect to a central A-atom or vice versa.

2.5.2 The radial distribution function

From these structure factors or partial structure factors we can derive by Fourier transforms the radial distribution functions appropriate to the alloy. These tell us the average number density of the different types of atom as we move radially outward from a given type of atom. An example is shown in Figure 2.6a. The positions of the peaks indicate the approximate distances of first, second, etc. neighbours from the atom chosen as origin, while the areas under the peaks indicate the numbers in the corresponding shells. The radial distribution function does not, however, give us any information about the angular positions of the neighbours. To that extent therefore there is a lack of detail.

From measurements on a number of metal–metalloid glasses we find that the number of nearest neighbours of the major component lies between 11.5 and 13 and the nearest-neighbour distance is about 12% greater than the effective diameter of the metal species. The partial structure factors of metal–metalloid alloys also reveal that in most of these alloys the metalloid ions, such as the boron in $Fe_{80}B_{20}$, are rarely nearest neighbours of each other, while the metal ions (here Fe) take up positions similar to those of a somewhat distorted close-packed crystal. Presumably the smaller metalloid atoms occupy the interstices of this structure.

This type of measurement confirms what direct measurements of the density show, that metallic glasses are densely packed materials. In general we find that, in terms of the fractional volume occupied by the atoms, the density of packing in metallic glasses is about 0.67, compared to about 0.64 for a liquid alloy and 0.74 for a face-centred cubic lattice.

The radial distribution function of a metallic glass looks very similar to that of the corresponding liquid (see Figure 2.6(b)), although there is a

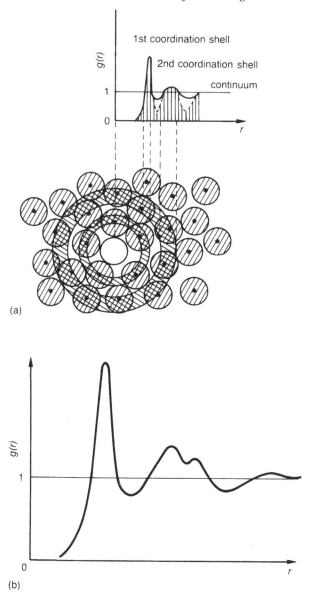

Fig. 2.6 Radial distribution function $g(r)$ of amorphous material. (a) Coordination shells; (b) radial distribution function. (After Ziman 1979.)

small systematic difference in the short-range order. This is seen as a splitting of the second peak for the glass as is shown in the figure; it has two shoulders instead of one as in the liquid.

Although the main source of short-range order is the close packing of the ion cores, another source of such order arises if there are directed bonding forces between the ions, which may then try to form molecules in the liquid; these near-molecules may persist into the glassy state.

Having seen briefly how metallic glasses are prepared and what we can tell about their structure, we turn now to a study of the conductivity and other electrical properties of metals to find out how far present theories can explain these and how the theory has to be extended to describe the properties of metallic glasses[1].

3

Electron transport in metals: introduction to conventional theory

3.1 The source of electrical resistance

In trying to understand the electron transport properties of metallic glasses – properties such as electrical conductivity, Hall coefficient and thermopower – we shall start by using conventional theories that have been successful in accounting for the corresponding properties in crystalline metals and alloys and see how far these theories are successful in describing the properties of metallic glasses. I will explain what I mean by 'conventional' theories as we go along.

The starting point in understanding the electrical conductivity, σ, or resistivity, $\rho(= 1/\sigma)$ of metals is the fact that the de Broglie waves which represent the conduction electrons can propagate without attenuation through a perfectly periodic lattice, such as that formed by the positive ions of an ideally pure and perfect crystalline metal at absolute zero. There is thus no electrical resistivity. More strictly, the electrons are scattered by the ions but only coherently as in Bragg reflections from the lattice planes. Such coherent scattering alters the way the electrons respond to applied electric and magnetic fields but does not cause electrical resistance. Such resistance comes about through random, incoherent scattering of the electron waves; this occurs only when the periodicity of the lattice and its associated potential is upset.

This means that, if you now add to your pure and perfect crystal chemical impurities randomly distributed, they will disturb the perfect periodicity and cause resistance to the flow of the electric current. Likewise, physical imperfections such as vacancies, dislocations or grain boundaries will produce electrical resistance even at the absolute zero. These imperfections and chemical impurities upset the perfect periodicity and so scatter the electrons that carry the electric current.

The same is true of the lattice vibrations that arise when the temperature is raised. These vibrations likewise destroy the perfect periodicity of the lattice and cause resistance. This time, however, the resistance depends on the temperature; usually the higher the temperature of the metal the greater its resistivity because the amplitude of the lattice vibrations increases and this increases the scattering. In quantum terms, we would say it is because the number of phonons increases with rising temperature. To sum up: departures from perfect periodicity of the potential through which the conduction electrons flow cause electrical resistance. This tells us at once that amorphous systems, whether liquid or glass, will, because of their disordered structure, tend to have high resistivity at all temperatures.

3.2 The conduction electrons

To understand electrical conduction in metals or alloys we have to understand the properties of the conduction electrons, that is, those electrons that actually carry the current. In some simple metals such as sodium or potassium, even in the liquid state, and even in some glasses composed of simple elements, these electrons behave very much as if they were free charged particles except that they are confined within the volume of the metal and subjected to occasional scattering processes. This may seem a remarkable state of affairs given the strong electrostatic Coulomb forces between the electrons themselves and between the electrons and the ions. Let us, however, accept this unlikely suggestion, see where it leads us and, later on, try to understand how it comes about.

Electrons are charged particles of spin 1/2: this means that they obey the Pauli exclusion principle and are subject to Fermi–Dirac statistics. Consider first the implications of this at the absolute zero of temperature. The kinetic energy levels of a free particle in a box are known from quantum mechanics. Each such level can accommodate two electrons of opposite spin; at the absolute zero all the lowest of these levels that are needed to accommodate the conduction levels are filled. If there are N conduction electrons per unit volume, then in unit volume of the metal, the lowest $N/2$ kinetic energy levels will be filled up as indicated in Figure 3.1. Above them the levels are completely empty. The energy which separates the two sets is called the Fermi level (strictly speaking, the Fermi level at absolute zero). If we calculate this energy for the conduction electrons in potassium, which has just one conduction electron per ion, we find that the energy is about 2 eV. If you bear in mind that, at room

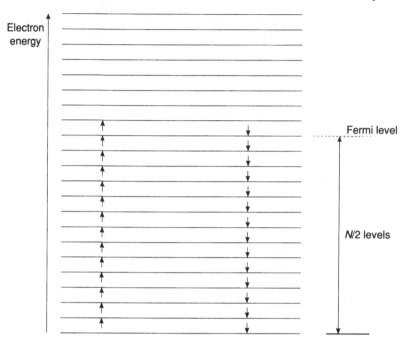

Fig. 3.1 Electrons in a metal at absolute zero; all levels are occupied by opposite spin pairs up to the Fermi level. Above that they are empty.

temperature, thermal energies are of the order of 1/40 eV, you see that the Fermi energy is a very large energy indeed, in this example, corresponding to the thermal energy of a classical gas at about 25 000 K. In most metals it is even higher. The important point, therefore, is that the electrons at the top of this distribution are very energetic with velocities of $10^6 \, \mathrm{ms}^{-1}$ or more, approaching 1% of that of light. Although non-relativistic these velocities are large and indeed go some way to explaining why these electrons can be treated as quasi-free particles.

3.2.1 The Fermi energy in a free-electron gas

If the kinetic energy of a particle of mass m is E and its momentum is p, then in classical terms:

$$E = p^2/2m \qquad (3.1)$$

From the de Broglie relationship, the wavelength associated with p is:

$$\lambda = h/p \qquad (3.2)$$

or, in terms of the wave vector $k = 2\pi/\lambda$:

$$p = \hbar k, \tag{3.3}$$

where h is Planck's constant and $\hbar$ is $h/2\pi$.

Thus

$$E = \hbar^2 k^2 / 2m \tag{3.4}$$

or, written in component form:

$$E = \hbar^2 (k_x^2 + k_y^2 + k_z^2)/2m \tag{3.5}$$

This shows that if we plot the constant energy surfaces as a function of k_x, k_y, k_z (this is called k-space) they form a set of concentric spheres about the origin. If the energy of the highest occupied level is E_0, then all the states inside the sphere corresponding to E_0 are full and all those outside it are empty. The surface dividing the occupied from the unoccupied states is called the Fermi surface.

The boundary conditions on the electron waves show that in k-space the allowed quantum states are uniformly distributed. Different boundary conditions lead to differences in detail but the same result. Here we take the periodic boundary conditions according to which the size of each quantum state in k-space is $(2\pi)^3/V$ where V is the volume of the metal. To find the number of k-states below the energy E_0, we must determine the number of such states within the sphere corresponding to this energy. This sphere has a radius k_F, the Fermi radius, such that:

$$\hbar^2 k_F^2 / 2m = E_0 \tag{3.6}$$

It therefore contains $4\pi k_F^3 V/3(2\pi)^3$ states. If this sphere is to accommodate N conduction electrons then, given that each k-state can hold two electrons of opposite spin, we require that:

$$N/2 = 4\pi k_F^3 V/3(2\pi)^3 \tag{3.7}$$

Thus,

$$k_F = (3\pi^2 N/V)^{1/3} \text{ and } E_0 = \hbar^2 (3\pi^2 N/V)^{2/3}/2m \tag{3.8}$$

So we see that if we know the density of conduction electrons (N/V) we can find E_0.

Some values of E_0 for a number of metals are given in Table 3.1. The energies are expressed in electron-volts (eV) and also as equivalent Fermi temperatures T_F where $k_B T_F = E_F$ (k_B is Boltzmann's constant). We see that these temperatures are in the range 10^4–10^5 K. Because the values of

Table 3.1   *Fermi energy and temperature, and density of states at E_0 for some metals.*

Metal	Fermi energy E_0 at 0 K (eV)	dn/dE (states per electron per eV)	Fermi temp. T_F (K)
Li	4.74	0.16	55,000
Na	3.16	0.24	37,000
K	2.06	0.36	24,000
Rb	1.79	0.42	21,000
Cs	1.53	0.49	18,000
Cu	7.10	0.106	82,000
Ag	5.52	0.134	64,000
Au	5.56	0.135	65,000

E_0 are so high compared to normal thermal energies, changing the temperature of a metal has very little effect on the Fermi distribution of energy among the electrons. Only those states within about $k_B T$ of the Fermi level (at E_0) are perturbed since only these have empty states into which their electrons can be excited. In lithium at room temperature, for example, this is only about 1% of the states. Moreover, it is only those electrons near to E_0 that can be scattered by the impurities, defects or lattice vibrations. The reason is the same, i.e. there are no other unoccupied states of comparable energy into which the electrons can be scattered. Bear in mind that the electric field that produces a current in a metal is of the order of a few volts per cm and imparts negligible energy of the electrons (about 10^{-4}–10^{-7} eV). Even if the electrons can pick up energy from the lattice this is either of order $k_B T$ or the maximum energy of a phonon. The latter is about $k_B \theta_D$, where θ_D is the Debye temperature of the lattice, typically around 100 to 400 K. Thus the so-called Fermi electrons, those within about $k_B T$ or $k_B \theta_D$ of the Fermi level, have an overriding importance in determining the electronic properties of the metal or alloy.

For this reason, we often need to know as much as possible about these electrons. One property of importance is the number of states per unit energy range to which they have access in the neighbourhood of E_0. This quantity is called the density of states at the Fermi level. We can easily calculate the density of states in k-space because the states are uniformly distributed there; then we convert to the density in terms of energy from the relationship between E and k.

The number of k-states in a sphere of radius k about the origin is, as before:

$$4\pi k^3 V/3(2\pi)^3$$

So the number of states lying between k and $k + \mathrm{d}k$ is:

$$\mathrm{d}n(k) = 4\pi V k^2 \mathrm{d}k/(2\pi)^3 \qquad (3.9)$$

(We ignore the spin states at this stage.)

To complete the calculation, we must now express this in terms of energy instead of k by using the relationship:

$$E = \hbar^2 k^2/2m \qquad (3.10)$$

or

$$k = (2mE/\hbar^2)^{1/2} \qquad (3.11)$$

Finally we get for each spin direction:

$$\mathrm{d}n(E)/\mathrm{d}E = V[(2m/\hbar^2)^{3/2} E^{1/2}/2\pi^2] \qquad (3.12)$$

This is the required density of states of a free-electron gas confined to a volume V. Some values of this quantity are listed in Table 3.1.

3.3 Calculation of electrical conductivity

In the conventional calculation of electrical conductivity, a number of approximations are made, which for the most part are valid for metals that are fairly pure with not too many defects and also for simple liquid metals and metallic glasses.

The most important of these approximations, the one-electron approximation, assumes that each conduction electron moves in the potential field of the ions and in the average field of all the other electrons. This means that each conduction electron experiences a potential that depends only on the ionic positions and on the coordinates of that electron alone; the potential does not depend on the distances between the electron of interest and all the other conduction electrons. This is a huge simplification but we shall assume its validity for the present and see where it leads us.

3.3.1 The distribution function

To describe the behaviour of a collection of conduction electrons, we use a distribution function $g(\mathbf{k}, \mathbf{r}) \, d^3k d^3r$ which tells us the number of electrons in a volume d^3r around the point $\mathbf{r}$ with $\mathbf{k}$-vectors lying in a volume d^3k around the value $\mathbf{k}$. In our case we shall consider only homogeneous metals at a uniform temperature so that in fact $g(\mathbf{k}, \mathbf{r})$ is independent of $\mathbf{r}$. Moreover it is convenient to use, not the number of particles with values around $\mathbf{k}$, but the fraction of the states that are occupied there. Since, as we saw, the states in k-space are uniformly distributed we can readily go from the fraction of states occupied to the number of electrons in the same volume of k-space by multiplying by the number of states per unit volume, which is just $V/(2\pi)^3$ (where V is the volume of the metal), or one-half of this when we take account of spin. Thus we write for the distribution function:

$$f(\mathbf{k}) = f(k_x, k_y, k_z) \, dk_x dk_y dk_z \qquad (3.13)$$

In the absence of applied fields and when we thus have an equilibrium distribution, the function f depends on $\mathbf{k}$ only through the energy E and is just the Fermi distribution:

$$f_0 = 1/[1 + \exp(E - E_F)/k_B T] \qquad (3.14)$$

Here E_F is the Fermi energy which at the absolute zero is the same as E_0 which we have already discussed.

3.3.2 The Boltzmann equation

We wish now to find out how the distribution function f changes when the metal and hence the electron gas is subjected to external influences such as electric or magnetic fields. To do this we follow the method that Boltzmann used in treating the properties of classical gases. We focus attention on a small region of phase space, here k-space, and ask how $f(\mathbf{k})$ at that point is influenced on the one hand by the applied fields and on the other by scattering processes which tend to restore the electron gas to its equilibrium state. Quite formally we write:

$$df/dt = (df/dt)_{\text{fields}} + (df/dt)_{\text{collisions}} \qquad (3.15)$$

to indicate the two quite different processes that are going on. We are concerned here only with steady-state conditions as they occur in, for example, the steady flow of a current through a metal wire. Thus the left-hand side of equation (3.15), which represents the total rate of change of f

with time, is zero. This means that the changes in f brought about by fields are just compensated by the scattering or collision processes. We now look at each of these in turn.

3.3.3 *The influence of fields*

Suppose that we apply a uniform electric field ϵ in the x-direction to the electrons in a metal in order to produce an electric current. The force on each electron is $e\epsilon$ where e is the charge on the electron. (We make it an algebraic quantity, which would in this context be negative.) The equation of motion of the electron with wave vector **k** we take to be of the same form as that of the free particle (the quantum analogue of Newton's second law with the rate of change of momentum written as $\hbar d\mathbf{k}/dt$):

$$\hbar dk_x/dt = e\epsilon \tag{3.16}$$

This implies that in a short time interval δt all the occupied states in k-space are displaced by an amount:

$$\delta k_x = e\epsilon\delta t/\hbar \tag{3.17}$$

Thus the new distribution f is the same as f_0 but displaced by δk_x. Thus:

$$f = f_0(k_x - \delta k_x, k_y, k_z) \tag{3.18}$$

Since we are concerned only with very small displacements in k-space (we assume in effect that Ohm's law is obeyed) we can expand this and take only the first term so that we have, making use of equation (3.17):

$$f = f_0 - (\partial f_0/\partial k_x)e\epsilon\delta t/\hbar \tag{3.19}$$

Since, however, f_0 depends on k_x only through E we can write:

$$\partial f_0/\partial k_x = (df_0/dE)(\partial E/\partial k_x) \tag{3.20}$$

or, since $\partial E/\partial k_x = \hbar v_x$, where v_x is the x-component of the electron velocity, we have:

$$f = f_0 - (df_0/dE)\hbar v_x(e\epsilon/\hbar)\delta t \tag{3.21}$$

Finally, therefore, we get:

$$(df/dt)_{\text{fields}} = -(df_0/dE)v_x e\epsilon \tag{3.22}$$

3.3.4 The influence of collisions or scattering

In considering the influence of scattering processes on the distribution function f, we must bear in mind that these processes go on all the time whether or not the electrons are in equilibrium. These scattering processes maintain equilibrium when there are no perturbations (other than temperature) and strive to restore equilibrium when it is disturbed.

We assume that at any point on the Fermi surface the rate of change of f due to scattering is given by:

$$(df(\mathbf{k})/dt)_{\text{collisions}} = -[f(\mathbf{k}) - f_0(\mathbf{k})]/\tau \qquad (3.23)$$

where f_0 is the equilibrium distribution function. This implies that if at any instant the actual distribution at a point $\mathbf{k}$ on the Fermi surface differs from that at equilibrium, the distribution will, if left to itself, return exponentially to equilibrium with a characteristic time τ.

Let us now put together equations (3.22,3.23) and the Boltzmann equation (3.15) so that we get:

$$-(df_0/dE)v_x e\epsilon - [f(\mathbf{k}) - f_0(\mathbf{k})]/\tau = 0 \qquad (3.24)$$

which on rearrangement gives:

$$f(\mathbf{k}) = f_0(\mathbf{k}) - (df_0/dE)v_x e\epsilon\tau \qquad (3.25)$$

This is the distribution function that describes the electron population under the combined influence of a steady electric field ϵ and random scattering processes characterised by a relaxation time τ. We can now use this to calculate the electrical conductivity of a metal.

3.4 The electrical conductivity

Let us apply an electric field ϵ_x in the x-direction of the metal and then calculate the resulting current density j_x in that direction. If v_x is the x-component of the velocity of an electron with charge e, then the current density is given by:

$$j_x = \Sigma e v_x \qquad (3.26)$$

where the sum is over all the conduction electrons in unit volume of the metal. It is, however, much easier to deal with an integral rather than a sum so the next step is to make this transition by means of the distribution function. This gives us the fraction of states occupied in any small region of k-space and we combine this with the fact that for unit volume of metal each electron state occupies a volume of $4\pi^3$ in k-space. Thus equation (3.26) can be rewritten:

$$j_x = (e/4\pi^3) \int f(\mathbf{k}) v_x \mathrm{d}^3 k \qquad (3.27)$$

where $\mathrm{d}^3 k$ stands for an elementary volume of k-space. We now use equation (3.25) for $f(\mathbf{k})$ and obtain:

$$j_x = (e/4\pi^3) \int [f_0 - v_x e \epsilon_x \tau (\mathrm{d}f_0/\mathrm{d}E)] v_x \mathrm{d}^3 k \qquad (3.28)$$

The integral has two parts, of which the first involves f_0; this contributes nothing since it describes a distribution that is completely symmetrical about the origin. The second involves $\mathrm{d}f_0/\mathrm{d}E$; this is a function which, for metals at temperatures with which we are normally concerned, is vanishingly small for all values of E except those in the immediate vicinity of E_0, that is to say, within about $\pm k_B T$ of E_0. It is shown schematically in Figure 3.2. This indicates that contributions to the integral come only from regions in k-space that hug the Fermi surface i.e. the constant energy surface at E_0. This suggests that we should choose our element of volume $\mathrm{d}^3 k$ in a special way. We write the volume element as $\delta S \delta k_n$ where δS is an element of area of a constant energy surface and δk_n is an element of length in k-space normal to δS. Thus the element of volume in k-space is a small cylinder lying on a constant energy surface with its axis normal to this surface, as pictured in Figure 3.3.

δk_n has the direction of the velocity of the electrons in its neighbourhood and can be expressed in terms of this velocity. If the two neighbouring constant energy surfaces linked by δk_n differ in energy by δE, we can write

$$\delta k_n = (\partial k_n/\partial E)\delta E \qquad (3.29)$$

where $\partial k_n/\partial E = 1/\hbar v$ and v is the electron group velocity at this point. Thus finally the volume element can be written as:

$$\mathrm{d}^3 k = (1/\hbar v)\mathrm{d}S\mathrm{d}E \qquad (3.30)$$

If we put this in equation (3.28) we get:

$$j_x = -(e^2 \epsilon_x/4\pi^3 \hbar) \int \int (\tau v_x^2/v)\mathrm{d}S(\mathrm{d}f_0/\mathrm{d}E)\mathrm{d}E \qquad (3.31)$$

where the first integral is over a surface of constant energy E and the second over all energies.

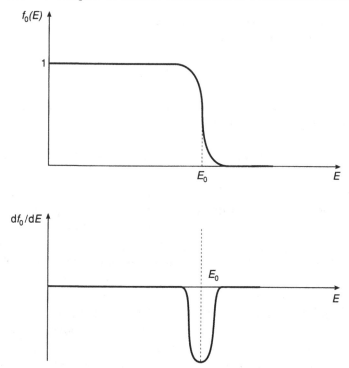

Fig. 3.2 The Fermi function and its derivative for a degenerate gas.

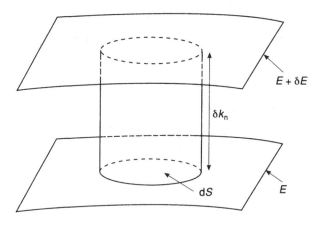

Fig. 3.3 Volume element between constant energy surfaces at E and $E + \delta E$ in k-space.

Since we shall almost always be dealing with metals in which the electron gas is highly degenerate, we can make an important simplification to equation (3.31): we can treat $-df_0/dE$ as a delta function, since it has the property that $\int_0^{\infty} -(df_0/dE)dE = f_0(0) - f_0(\infty) = 1$. This implies that only the surface at E_0 contributes to the integral i.e. that only electrons on the Fermi surface determine the conductivity. So finally we have:

$$j_x = (e^2/4\pi^3\hbar)\epsilon_x\tau \int (v_x^2/v)\mathrm{d}S \qquad (3.32)$$

where the integral is over the Fermi surface. In taking τ outside the integral we are assuming it is the same over the whole surface, an assumption valid for isotropic systems such as glasses but often not valid for crystals.

Again, if we confine ourselves to isotropic systems, then we know that:

$$v_x^2 = v_y^2 = v_z^2 = v^2/3 \qquad (3.33)$$

so that more generally equation (3.32) becomes:

$$j = (e^2/12\pi^3\hbar)\epsilon\tau \int v\mathrm{d}S \qquad (3.34)$$

If we compare this with Ohm's law $j = \sigma\epsilon$ we see that the conductivity is:

$$\sigma = (e^2\tau/12\pi^3\hbar) \int v\mathrm{d}S \qquad (3.35)$$

In cubic and, of course, isotropic metals σ is a scalar quantity.

If we now apply this to conduction by isotropically distributed electrons, we can put for their velocity at the Fermi level: $mv = \hbar k_F$ where v is the same over the whole surface. Moreover $\int \mathrm{d}S = 4\pi k_F^2$ and if we also put for the number of electrons in states inside the Fermi surface $n = 4\pi k_F^3/3(4\pi^3)$, we get:

$$\sigma = ne^2\tau/m \qquad (3.36)$$

which is just the simple Drude expression for σ.

Almost everything that we have derived here can be applied to amorphous metals, in particular equations (3.31), (3.35) and sometimes (3.36) provided that the mean free path of the electrons is large compared to their wavelength at the Fermi level. There are some many-body and multiple-scattering effects that we shall need to introduce but the basic ideas considered so far will take us some way in our understanding of metallic glasses.

3.5 Conductivity and diffusion coefficient of conduction electrons

There is another way of writing the conductivity of a metal that will be useful to us later. We start with equation (3.28) and, because the first term in the integral contributes nothing, we rewrite it as:

$$j_x = -(e^2\epsilon/4\pi^3) \int \tau (\mathrm{d}f_0/\mathrm{d}E) v_x^2 \mathrm{d}^3 k \qquad (3.37)$$

We now use equation (3.33) to replace v_x^2 by $v^2/3$. We also exploit the isotropy of the electron properties to replace $\mathrm{d}^3 k/4\pi^3$, which is just the number of conduction electron states in the volume element $\mathrm{d}^3 k$, by $N(E)\mathrm{d}E$. $N(E)$ is the density of such states with energy E and so $N(E)\mathrm{d}E$ is the number of states with energies between E and $E + \mathrm{d}E$. Because τ, $\mathrm{d}f_0/\mathrm{d}E$ and v depend only on E, we can in this way replace the integral over $\mathbf{k}$ with a more convenient integral over E. We now have:

$$j = -e^2\epsilon \int (\tau v^2/3)(\mathrm{d}f_0/\mathrm{d}E)N(E)\mathrm{d}E \qquad (3.38)$$

where I have dropped the x-subscripts because the metal is isotropic. If now we use the fact that the function $\mathrm{d}f_0/\mathrm{d}E$ approximates to a δ-function at the Fermi level, we can carry out the integration and find for the conductivity:

$$\sigma = e^2(\tau v^2/3)N(E_F) \qquad (3.39)$$

where the density of states is now that appropriate to the Fermi energy. Finally, we know from kinetic theory that the diffusion coefficient D of a gas of non-interacting particles with velocity v and mean free time between collisions τ is given by:

$$D = v^2\tau/3 \qquad (3.40)$$

so that equation (3.39) can be written:

$$\sigma = e^2 D N(E_F) \qquad (3.41)$$

This expression, often called the Einstein relation, tells us that the conductivity of a metal or alloy with isotropic properties is proportional to the density of states at the Fermi level and to the diffusion coefficient of the Fermi electrons; it is frequently used in discussions of the conductivity of metallic glasses.

Equation (3.40) is usually derived for a gas of particles scattering off each other but it applies equally well if the particles are scattered by other independent scatterers, which thus determine the mean free path of the gas particles. This corresponds to the conditions in which we shall apply it. The derivation implies that the particles are independent, that the material is isotropic and that the scattering processes randomise the directions of the scattered particles. This last condition in turn implies that τ refers to what has been called 'catastrophic' scattering i.e. the electron loses all memory of its past history. This is consistent with its significance in equation (3.39) for the electrical conductivity.

In all the derivations based on the Boltzmann equation, we are assuming that the electrons follow semi-classical paths in the sense that there is no interference between the wave functions at different places along their path. This is valid provided that the mean free path, l, between scattering events is very long compared to the electron wavelength. When this is no longer true, new and important features come into the story.

4

Scattering

The results that we have just obtained are rather formal but do give us some important insights into the nature of electrical conduction in metals. We see from equation (3.35) that this conductivity depends entirely on the properties of the conduction electrons at the Fermi level. Moreover, the properties involved are of two distinct kinds: the first kind relates to the dynamics of the electrons, as represented by the distribution of electron velocities over the Fermi surface. The second kind represented by τ is concerned with scattering, our theme in this chapter.

As we have already noted, scattering in a metal arises from anything that upsets the periodicity of the potential: disorder of the ionic positions, which is paramount in metallic glasses; random changes in chemical composition, which are of great importance in random alloys; impurities, physical imperfections, thermal vibrations, random magnetic perturbations and so on. Let us therefore see how the scattering from some of these can be treated.

In this chapter we deal in section 4.1 with some basic ideas about scattering theory; in 4.2 there is a very brief discussion of Fourier transforms because they appear so frequently in scattering problems; in 4.3 we look at the influence of scattering angle on resistivity; in 4.4 at the effect of the Pauli exclusion principle on scattering; in 4.5 we consider electron screening in metals because the mobile electrons can markedly alter the scattering potential by electrical screening of the scatterer; and in 4.6 the pseudopotential because this has been used very successfully to represent the scattering potential in some simple amorphous metals.

4.1 Some definitions and formulae of scattering theory

Suppose that a beam of electrons all with velocity v falls on a set of fixed scattering centres, N per unit volume, with scattering cross-section σ and sufficiently far apart that they scatter independently. The mean free path for scattering is then $1/N\sigma$ and the mean time between scattering events is $\tau = 1/N\sigma v$. σ refers to the total scattering cross-section, integrated over all scattering angles. We shall need to pay attention to the angle through which an electron is scattered and so the differential cross-section is important for our purposes. Thus we think of the scattering into an elementary solid angle $d\Omega$ at polar angles θ, ϕ to the direction of the beam, the z-direction, say. The differential cross-section is then $d\sigma(\theta, \phi)/d\Omega$; in general we assume that the scattering potential has spherical symmetry so that there is no dependence on ϕ (see Figure 4.1).

We now consider in outline how to calculate the cross-section for scattering of free electrons from a single scattering centre whose potential is represented by $V(r)$ where r is measured from the centre of the scatterer. The range of the potential is a so that $V = 0$ when $r > a$. This means that the scattered particles or waves then have the same energy or frequency as before they were scattered.

Let us assume that the electron in the presence of the scatterer has a wavefunction $\psi_\mathbf{k}(\mathbf{r})$ and we wish first to know what is the probability that it will be scattered into a specific state $\phi_{\mathbf{k}'}$ in which we are interested. The scattering probability per unit time is then given by Fermi's golden rule:

$$P_{\mathbf{k}'\mathbf{k}} = (2\pi/\hbar)|\langle \phi_{\mathbf{k}'}^*|V|\psi_\mathbf{k}\rangle|^2 N(E_0) \qquad (4.1)$$

Here $\langle \phi_{\mathbf{k}'}^*|V|\psi_\mathbf{k}\rangle$ is the matrix element of the scattering potential between the two states involved. $N(E_0)$ is the density of states at energy E_0 into which the electrons can be scattered.

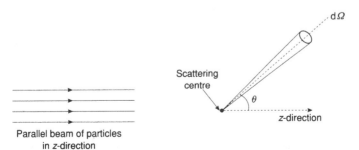

Fig. 4.1 Scattering of particles from a spherically symmetrical scattering centre. The particles are scattered through an angle θ into an element of solid angle $d\Omega$.

To proceed further, we make a substantial approximation, which is commonly made, although better approximations are now well developed and often used. We replace the initial state wavefunction $\psi_\mathbf{k}$ by $\phi_\mathbf{k}$, the wavefunction of the electron in the absence of the scattering centre. This approximation, known as the first Born approximation, is valid if $(V/\frac{1}{2}mv^2)(ka) \ll 1$. I have written the condition in this form to bring out the physics implied by the restriction. For the first bracket to be small, the kinetic energy of the electron must be large compared to the strength V of the scattering potential; for the second bracket to be small, the wavelength of the electron must be large compared to a, the range of the scattering potential. Thus if one bracket amply satisfies the condition it takes the pressure off the other. In practice we usually think of the Born approximation as useful for particles of relatively high energy. Moreover if we are permitted to treat the electrons as free particles, the unnormalised wavefunctions corresponding to k and k' are $\exp(i\mathbf{k} \cdot \mathbf{r})$ and $\exp(i\mathbf{k}' \cdot \mathbf{r})$; we can omit the time-dependent factor because it is unchanged in this elastic scattering process. Thus the matrix element in equation (4.1) becomes:

$$V_{\mathbf{kk'}} = \langle \psi_{\mathbf{k}'}^* | V(r) | \psi_\mathbf{k} \rangle = \int [\exp(-i\mathbf{k}' \cdot \mathbf{r})] V [(\exp(i\mathbf{k} \cdot \mathbf{r})] \mathrm{d}^3 r \qquad (4.2)$$

where $\mathrm{d}^3 r$ is an element of volume and the integration is over all space. We now write $\mathbf{k}' - \mathbf{k} = \mathbf{K}$ where $\mathbf{K}$ is called the scattering vector. Equation (4.2) can then be written as:

$$V_{\mathbf{kk'}} = \int V(r) \exp(i\mathbf{K} \cdot \mathbf{r}) \mathrm{d}^3 r \qquad (4.2a)$$

which is the Fourier transform of V. It means that the scattering from $\mathbf{k}$ to $\mathbf{k}'$ involves that Fourier component $\mathbf{q}$ of the scattering potential which is just equal to the scattering vector $\mathbf{K}$ i.e. the difference between the outgoing and incident wave vectors. This in turn means that the prominent wave vectors in the scattered waves will be those corresponding to the strong Fourier components in the scattering potential. This seems physically very reasonable. Indeed we encountered a similar result in our earlier discussion of the determination of the structure of metallic glasses by scattering experiments and we shall frequently find that Fourier components and transforms provide a convenient language in which to describe the results of scattering processes. Let us therefore familiarise ourselves a little with these ideas before going further.

4.2 Fourier transforms

Electrons are described in terms of their wavefunctions whose important features are their frequency ω and wave vector $\mathbf{k}$. The wave vector is as its name implies a vector, which can therefore be resolved into components along directions of interest. That is why it is used in preference to the wavelength, which is more readily visualised but does not have this desirable property. That is also why k-space or reciprocal space plays such a prominent role in discussions of electrons in metals.

One of the valuable features of Fourier transforms is that it enables us to express some quantity (electrostatic potential, say) that varies from place to place in real space in terms of a set of periodic variations that can be compared directly with those of the electrons of interest. This sort of analysis is familiar in sound. A long note of a single frequency produces a sinusoidal variation in air pressure; this is one method of describing its properties, in this case its behaviour in real time. Alternatively it can be described by its frequency spectrum which here consists of a large spike at the signal frequency together with some smaller amplitude at other frequencies associated with the beginning and ending of the sound wave. More complex sounds will give rise to more complex spectra. Just as an infinitely long signal of a single frequency produces a delta function spectrum at that frequency, so reciprocally an infinitely sharp spike of signal in real time has a frequency spectrum involving a uniform distribution of the full (infinite) range of frequencies. Representations of variations in real space are analogous. The Fourier transform of a delta function in real (or direct) space is a constant distribution of wave numbers in reciprocal space.

A central feature of the study of electrons in crystals is the crystal lattice itself. This, in its ideal form, is a three-dimensional periodic array of points filling all space. Its Fourier transform is another periodic array of points forming a lattice in reciprocal space. This so-called reciprocal lattice represents all the periodicities that are present in the direct lattice.

If we wish to determine what waves will be scattered elastically from such a crystal lattice we first represent the incident wave by a vector $\mathbf{k}$ drawn from one of the reciprocal lattice points chosen as origin; then we draw from its tip another vector of equal length to represent the scattered wave $\mathbf{k}'$ (see Figure 4.2). This represents a possible scattering process only if the resultant vector $\mathbf{K} = \mathbf{k}' - \mathbf{k}$ lands on another reciprocal lattice point (Figure 4.2(b)). When this happens, it defines a possible Bragg reflection.

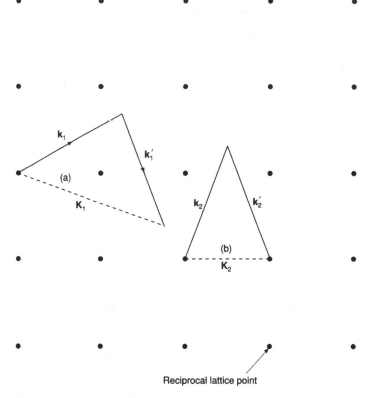

Fig. 4.2 Reciprocal lattice of a crystal. (a) No scattering possible because the vector $k + k'$ does not end on a reciprocal lattice point. (b) Possible scattering process because $k + k'$ does so end.

If it lands anywhere else it means that there is no Fourier component of the lattice potential with which to interact and so no scattering can occur (Figure 4.2(a)).

We shall often be concerned with the Coulomb interaction between electrons and between electrons and ions. In a metal the conduction electrons and positive ions screen the interaction and produce a screened Coulomb potential of the form:

$$V(r) = e^2 \exp(-\alpha r)/\epsilon_0 r \tag{4.3a}$$

where α is the reciprocal of the screening radius. The Fourier transform of this function, which will be useful to us later is:

$$V(q) = e^2/\epsilon_0(q^2 + \alpha^2) \tag{4.3b}$$

4.2.1 Better scattering approximations

There are higher approximations in scattering theory than the first Born approximation and, as the name implies, we can improve on the first approximation by using higher-order approximations to the incident wave. Another technique is known as phase shift analysis. In this the incident and outgoing waves are spherical waves and since we are assuming a spherically symmetrical scattering centre, there is conservation of angular momentum in addition to that of energy and number of particles. The consequence is that the only property of the wave that can change on scattering is its phase. The analysis gives the phase shifts of the different angular momentum components, which are labelled s,p,d,f, . . . according as the particle has zero, one, two, three . . . units of angular momentum about the scattering centre expressed in terms of $\hbar$. This is often a convenient way of describing the scattering especially when one angular momentum component is known to be dominant.

4.3 The influence of scattering angle

The effect of a scattering process on the electric current must depend on the angle through which the electron is scattered. Suppose that we apply to the metal an electric field in the x-direction. The contribution to the electric current of a given electron is proportional to the x-component of its velocity and so when it is scattered the change in its contribution is proportional to the change in this component. Thus:

$$\Delta j \propto (v_x' - v_x) \tag{4.4}$$

where the dash indicates the quantity after scattering. Moreover if the electrons can be treated as quasi-free as is often true in simple metals then their velocities are proportional to their wave vectors so we can write:

$$\Delta j \propto (k_x' - k_x) \tag{4.5}$$

where k_x and k_x' are the x components of the k-vector before and after scattering.

If we take the special case of an electron initially moving in the x-direction, the change in its x-component of k on being scattered through an angle θ is (see Figure 4.3):

$$\Delta k = k_F(1 - \cos \theta) \tag{4.6}$$

Moreover this turns out to be quite generally true of electrons on a spherical Fermi surface so that if the probability of scattering from

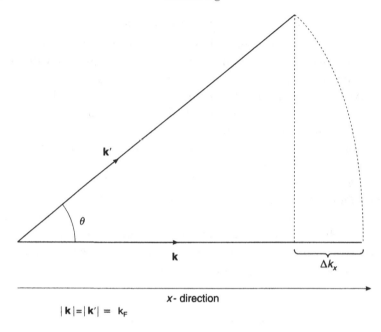

$$|\mathbf{k}| = |\mathbf{k}'| = k_F$$

Fig. 4.3 The change of k on scattering: $\Delta k_x = k_F(1 - \cos\theta)$.

state $\mathbf{k}$ to state $\mathbf{k}'$ is $P(\theta)$, which depends only on θ, the angle between $\mathbf{k}$ and $\mathbf{k}'$, then the relaxation time τ appropriate to electrical conductivity is given by:

$$1/\tau = P(\theta)(1 - \cos\theta) \tag{4.7}$$

This result can be expressed in terms of the differential cross-section for scattering by using the relation:

$$P(\theta) = v\,\mathrm{d}\sigma/\mathrm{d}\Omega \tag{4.8}$$

Equation (4.7) shows that, as we would expect, large-angle scattering is much more effective in reducing the current than small-angle scattering. This is particularly important when we come to consider scattering by phonons.

4.4 The influence of the exclusion principle

Conduction electrons are subject to the Pauli exclusion principle, which we have so far ignored in our discussion of scattering. At first sight we might think that as the temperature rose the spread of states into which electrons could be scattered would increase in proportion to the temperature and that the resistance would rise correspondingly even though the scattering didn't change. But we would be wrong. Let us first consider scattering processes that take electrons out of the state $\mathbf{k}$ into state $\mathbf{k}'$. This involves the probability that $\mathbf{k}$ is occupied, the probability that $\mathbf{k}'$ is empty and the probability $P(\mathbf{k},\mathbf{k}')$ of a transition between them. Then we must consider the inverse scattering processes that take electrons from state $\mathbf{k}'$ into $\mathbf{k}$. This involves the probability that $\mathbf{k}'$ is occupied, the probability that $\mathbf{k}$ is empty and the probability $P(\mathbf{k}',\mathbf{k})$ of a transition between them. Because of the requirement of detailed balance at equilibrium, $P(\mathbf{k},\mathbf{k}')$ must equal $P(\mathbf{k}',\mathbf{k})$ and so finally it turns out that the expression for the rate of change of the population of any particular k-state is exactly what we would have got if we had ignored the exclusion principle. That is why if the scattering cross-section is independent of temperature the associated resistivity is likewise independent of temperature.

4.5 Electron screening in metals

One important difference between an isolated scattering centre and one in a metal is that in the metal the conduction electrons, being mobile, can move to screen the field of the scattering centre. Let us first consider a simple example, Thomas–Fermi screening, which has application to metallic glasses.

An impurity is introduced into the metal so replacing an ion of the host metal by one whose charge differs from that of the host by Ze where e is the magnitude of the electronic charge and Z is the valence difference between impurity and host.

Figure 4.4(a) shows the filled states and the potential energy of an electron at the bottom of the band in the metal before the impurity is introduced; in this approximation it is everywhere constant. The positive ions of the pure metal are represented by a uniform positive charge of $N_0 e$ per unit volume to compensate exactly for the negative charge on the N_0 conduction electrons per unit volume. Figure 4.4(b) shows how the potential energy at the bottom of the band is changed when we introduce the impurity. If Z is positive, the impurity introduces an excess of electrons which tend to stay in its neighbourhood filling up the region of low potential around it shown in the figure. (If Z is negative, the electrons are correspondingly repelled.) Close to the impurity the change in potential $\delta\phi$ is just the bare Coulomb potential $Ze/4\pi\epsilon_0 r$ where r is the distance

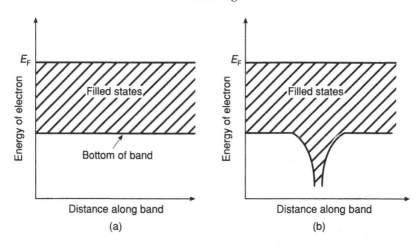

Fig. 4.4 Occupied energy states inside metal (a) before and (b) after introducing an impurity. Note the change in energy of an electron at the bottom of the band corresponding to the change in potential there.

from the ion. At large distances, because of screening, the change in potential at the bottom of the band goes to zero, as indicated in the figure.

To find the form of the screened potential $\delta\phi$ we use Poisson's equation:

$$\nabla^2\delta\phi = -\rho/\epsilon_0 = -e[N_0 - N(r)]/\epsilon_0 \qquad (4.9)$$

where ρ is the charge density and $N(r)$ is the number of electrons per unit volume at a distance r from the impurity. The second term in the bracket gives the local density of charge due to the electrons and the first term that of the uniform positive background due to the ions, assumed to be fixed.

We now make use of the fact that if the electrons are in equilibrium the Fermi level, which represents the chemical potential of the conduction electrons, remains constant everywhere. This implies that where the potential energy of the electrons is reduced in the neighbourhood of the ion their kinetic energy must increase to compensate for it. The required kinetic energy is related to the local density of electrons and so allows us to express $N(r)$ in terms of the change in the local electrostatic potential $\delta\phi$ that appears in Poisson's equation. For simplicity and as a good approximation in the circumstances where we shall employ it, we assume that the metal is at absolute zero. The argument goes as follows. Since the potential at a distance r from the impurity differs from that without the impurity by an amount $\delta\phi$, the corresponding energy change is $e\delta\phi$ and so the number of electron states required to compensate for this is:

$$\delta n = D(E_F)e\delta\phi \qquad (4.10)$$

where $D(E_F)$ is the density of states at the Fermi level. (We use D instead of N to avoid confusion with N the number density of electrons.) Thus in equation (4.9), $N(r) - N_0 = \delta n = D(E_0)e\delta\phi$ so that we get:

$$\nabla^2\delta\phi = e^2 D(E_0)\delta\phi/\epsilon_0 \qquad (4.11)$$

The solution to this equation that satisfies the boundary conditions that we specified above is:

$$\delta\phi = +Ze\exp(-\chi r)/4\pi\epsilon_0 r \qquad (4.12)$$

where $$\chi^2 = e^2 D(E_0)/\epsilon_0 \qquad (4.13)$$

χ has the dimensions of reciprocal length and $1/\chi$ is called the screening radius; it is a rough measure of the effective range of the screened potential. The bare Coulomb potential with which we started is now exponentially damped. Thus, as we see from equation (4.13), if there is a high density of mobile electrons and so a high density of states at the Fermi level the screening is strong and the effective radius small and vice versa.

The Fourier transform of equation (4.12) is:

$$\delta\phi(q) = Ze/D(E_0)[q^2 + \chi^2] \qquad (4.12a)$$

Notice that this increases as q tends to zero, whereas according to equation (4.12) $\delta\phi(r)$ goes to zero as r tends to infinity. This behaviour as a function of q is important to us later and we shall there discuss its significance in more detail.

This result shows how we can take account of electron screening of a single scattering centre. There are better approximations and the one most often used for the ionic potential in simple metals is the Lindhard screening function. It acts in a manner analogous to the dielectric constant in an insulator. If the Fourier component corresponding to the wave number q of the bare (unscreened) potential is $V(q)$, the screened potential can be written as:

$$U(q) = V(q)/\epsilon(q) \qquad (4.14)$$

where

$$\epsilon(q) = 1 + (3\pi^2 N_0/q^2 E_F)\{1 + [(1 - \beta^2)/2\beta]\ln|(1 + \beta)/(1 - \beta)|\} \quad (4.15)$$

Here $\beta = q/2k_F$. Equation (4.15) is the Lindhard function for screening by a highly degenerate gas consisting of N_0 quasi-free electrons per unit volume, whose Fermi energy is E_F and Fermi wave vector k_F.

Without going into any detail, we can understand some of the physics underlying this expression as follows. The component q of the screening function comes into play when an electron is scattered from $\mathbf{k}$ to $\mathbf{k}'$ (both on the Fermi surface) so that $\mathbf{q} = \mathbf{K} = \mathbf{k}' - \mathbf{k}$ where $\mathbf{K}$ is the scattering vector. This at once tells us that q must lie between 0 and $2k_F$. This also implies that the electron's ability to screen depends on the wavelength that results from the scattering. For example the shortest wavelength available arises from the scattering of an electron diametrically across the Fermi sphere, which produces a value of $2k_F$ for the wave number. This therefore sets a limit to the perfection of screening that can be achieved by the conduction electrons and leaves a residual ripple in the screening charge. Such ripples are referred to as Friedel oscillations and are implicit in equation (4.15). At the long-wavelength end (i.e. as q tends to zero and therefore describes features at large distances) the expression becomes the same as that for Thomas–Fermi screening. In this limit as we saw above the screening gets better and better and ultimately becomes perfect, thereby cutting off any long-range effects of the unscreened Coulomb force.

Equation (4.15) implies that the k-states of the electrons are well defined and can satisfy the above relation between $\mathbf{k}$, $\mathbf{k}'$ and $\mathbf{q}$. In a disordered system, however, the wave trains that represent the electrons are limited by scattering to an average length of l, the mean free path, and this produces uncertainties in k and k' of order $1/l$. Consequently if $q < 1/l$, q becomes meaningless and under these conditions the disorder will upset the screening, an effect we shall discuss in more detail below.

4.6 The pseudopotential[1]

For non-transition metals the so-called pseudopotential provides a comparatively simple way of describing the potential of an ion as seen by a conduction electron. The great problem of this electron–ion potential is that at distances close to the nucleus it becomes very large and ultimately diverges. The situation is saved, however, because inside the ion where this happens, there are the so-called core electrons which form completely closed shells and through the operation of the Pauli exclusion principle these repel the conduction electrons that seek to intrude. They provide an effective repulsive potential inside the ion and also partly screen the conduction electrons from the full nuclear charge. Outside the ion core we would thus expect the potential to be the Coulomb potential of a point charge Ze (we assume that the ion is spherical) where Z is the valence of the ion i.e. the number of electrons lying outside the closed shells. This potential is further screened by the conduction electrons which, as we have seen, can move very rapidly into regions of low potential.

Strictly, the effect of the core electrons should be represented by making the conduction electron wavefunctions orthogonal to the ion core states. Alternatively the potential can be made into an operator acting on free electron wavefunctions to achieve the same effect. In some versions of the pseudopotential method, however, the operator is treated as an algebraic quantity and its effect is simulated by replacing the ion core by a repulsive term in the potential.

The combination of the repulsive term and the screening by the conduction electrons has the effect of removing the troublesome divergent inner potential and making the electron–ion pseudopotential weak enough that perturbation methods can be used in calculations. This makes the calculation of the necessary matrix elements straightforward and justifies the application of the nearly-free-electron model to many non-transition metals and alloys in crystalline, liquid and glassy form.

In discussing the pseudopotential, as in our discussion of scattering probabilities, it is convenient to work in terms of its Fourier transform. Thus we write it as follows:

$$U(q) = [-(Ze^2/\epsilon_0 q^2) + A]/\Omega_0 \epsilon(q) \qquad (4.16)$$

where Ω_0 is the volume per atom, A is the repulsive term and $\epsilon(q)$ is the Lindhard screening function already discussed in section 4.5.

The first term in the square bracket is the Fourier transform of the Coulomb potential of the ionic charge Ze; this is of long range and so dominates at small values of q. The repulsive part of the potential can take many forms; here it is represented in real space by a delta function of strength A at the origin, whose Fourier transform is just the constant A. This form was used by Harrison, who has tabulated the value of A (he uses the symbol β) for many ions. This repulsive term is positive and dominates at short range i.e. at large values of q.

The screening by the conduction electrons is described by $\epsilon(q)$ which as we saw earlier acts as a dielectric function. As $q \to 0$ we can use Thomas–Fermi screening as given in equation (4.12a) and write equation (4.16) as:

$$U(q) = -[Ze^2/\epsilon_0 q^2][q^2/(q^2 + \alpha^2)] \qquad (4.17)$$

with $\alpha^2 = N(E_0)Ze^2/\epsilon_0$ and $N(E_0)$ the density of states. In the limit of $q = 0$, $U(q)$ becomes simply $-1/N(E_0)$. This is now constant so that there is no electric field and the screening is perfect. Moreover, the value of the pseudopotential in this limit has become independent of the character of the ion, a very remarkable result. One way to understand this apparently strange consequence of equation (4.17) is to recognise that the Fourier transform of a potential is closely related to the diffraction pattern formed by waves scattered from that potential (here the screened Coulomb potential). The component q is then the scattering vector of the diffraction pattern and, as we saw earlier, is related to the scattering angle θ; for Fermi electrons $q = 2k_F \sin\theta$. Therefore $q = 0$ corresponds to those waves that pass through the scattering centre undeflected. They are analogous to the central maximum of the diffraction pattern of light falling normally on a slit. The light that contributes passes through without any change of phase. Here the details are different but the principle is the same. We shall come across this effect again later.

The great advantage of the pseudopotential is that for a given density of screening electrons it is quite independent of its surroundings so that for a given ion it can be used when the ion is an impurity or part of the host metal or the constituent of an alloy. The screening function ϵ is calculated for free electrons at a density appropriate to the host metal or alloy.

The main features of the pseudopotential are illustrated in Figure 4.5 where you see that it begins at $q = 0$ large and negative, changes sign in the region of $q = 2k_F$ and continues positive, although it may thereafter change sign again. These features have important implications for the electrical properties of amorphous metals as we shall see.

The pseudopotentials are a valuable approximation but they have important limitations. They are not unique; they depend on the energy of the electron concerned, although this can often be ignored in the energy range for which they are chosen and used; the formalism is not valid when electrons from incomplete inner shells such as d-electrons are also involved. Nonetheless these potentials can be used to correlate many different properties of non-transition metals, in particular, electron transport, the band structure and Fermi surface, defects and cohesion. Often, for example, the empirical Fermi surface in the crystal is used to fix the parameters of the pseudopotential, which can then be used for other calculations, notably those relating to the amorphous state.

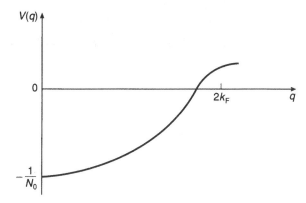

Fig. 4.5 The main features of the Fourier transform of the pseudopotential (schematic).

In the next chapter all the diverse elements of this and the preceding chapter come together in the Ziman theory of simple liquid metals to show that, at least in some examples, we can advance a long way in accounting for the electrical conductivity of amorphous metals.

5

Simple liquid metals: Ziman theory

5.1 k-states in disordered metals

The simplest amorphous metals are probably the liquid non-transition metals such as liquid sodium or liquid zinc. The simplicity arises partly because no d-electrons are involved and partly because the liquids are of a single component whereas by contrast all metallic glasses involve at least two components. So let us see how far we can understand the electron transport of these amorphous metals before we tackle systems with the additional complication of two or more components.

As soon as we confront the problem of electron transport in highly disordered systems like liquids or glasses several questions spring to mind. How useful is the concept of a k-state when we are so far from having translational symmetry? How valid is the concept of a Fermi surface? The answers depend, not surprisingly, on the degree of scattering involved. Thus it is not just the degree of disorder involved but also the strength of the individual scattering processes. If the mean free path of the electrons is l and the electron wavelength at the Fermi level is λ we require that l be much greater than λ. More commonly we choose the almost equivalent condition $k_F l \gg 1$ where $k_F = 2\pi/\lambda$. In a highly disordered system the mean free path tends to be only weakly temperature dependent so that the condition itself is essentially temperature independent. Experience with other systems of limited mean free path, for example, calculations on concentrated random binary crystalline alloys which give results in accord with experiment, suggests that a Fermi surface and its associated k-vectors for the conduction electrons are useful and satisfactory concepts provided that $k_F l \gg 1$. From the conductivity of liquid sodium, for example, we can infer that at its melting point $k_F l$ is of the order of 1000, so amply satisfying the above criterion.

When an electron is heavily scattered its quantum state can no longer be represented by unique values of its energy E and wave vector $\mathbf{k}$. Instead the range of values of E and k (in a particular direction) that are needed is shown by the 'spectral function' of the electron. Thus the E–k curve is no longer a two-dimensional graph but becomes a three-dimensional picture. The dispersion curve now looks like a range of hills rising above the E–k plane with the height above the plane at any point representing the probability density that the electron should have that value of E and k. The cross-section through the mountain range parallel to the E-axis shows the spread of k values needed to represent the state at that value of E and the cross-section parallel to the k-axis the spread of E values. If there were no scattering the E–k curve would be a line of δ-functions.

5.2 The Ziman model

The low field Hall coefficient of many of these metals is free-electron-like i.e. has the value $R_H = 1/ne$ where n is the number of conduction electrons per unit volume (see Chapter 9). This suggests that as a good first approximation we can treat these liquid metals as quasi-free-electron metals with spherical Fermi surfaces and an electrical conductivity σ given by the simple Drude-type formula:

$$\sigma = ne^2\tau/m \quad \text{or} \quad \rho = m/ne^2\tau \tag{5.1}$$

where m is an appropriate mass, which we take to be the free-electron mass, and $1/\tau$ is the scattering probability. For our purposes, it is more convenient to use the equivalent expression (3.39):

$$\sigma = e^2v^2N(E)\tau/3 \quad \text{with} \quad \rho = 1/\sigma \tag{5.2}$$

If this simple idea is correct the problem of calculating ρ resolves itself into that of calculating τ. This raises two questions: first, how do we calculate and describe the structure of the liquid? This is a question not yet fully answered. Second, how do we calculate the diffraction of the electron waves from this amorphous and hence rather complex structure. Ziman in 1961 suggested a way of answering both these questions simultaneously by use of the structure factor of the liquid. If we know this either from neutron or X-ray experiments or by calculation, we can use this directly to calculate the probability of electron scattering. As we saw in Chapter 2, the structure factor gives the relative intensity of the waves scattered from the target (here the liquid metal) when plane waves are incident on it (here the conduction electrons); the intensity is given as a function of scattering angle. If therefore we combine this information with the scattering cross-section of an individual ion of the liquid, we can get the total scattering rate of the electrons.

As we saw earlier, the structure factor is usually expressed in terms of the scattering vector $\mathbf{K}$ defined in terms of the wave vectors $\mathbf{k}$ and $\mathbf{k}'$ before and after scattering

$$\mathbf{K} = \mathbf{k}' - \mathbf{k} \qquad (5.3)$$

For electrons on a spherical Fermi surface

$$K = 2k_F \sin(\theta/2) \qquad (5.4)$$

where θ is the angle between $\mathbf{k}$ and $\mathbf{k}'$ and the radius of the Fermi surface is k_F. The maximum value of the scattering vector $\mathbf{K}$ is thus $2k_F$ i.e. when the scattering is diametrically across the Fermi sphere.

On the basis of these ideas we now wish to calculate the probability that a conduction electron in a simple liquid metal is scattered. We start from equation (4.1) and convert it into the probability per unit time $(1/\tau)$ that an electron with velocity v_F is scattered into a solid angle $d\Omega$. We also take account of the influence of scattering angle θ on resistivity with the factor $(1 - \cos\theta)$. Thus we find:

$$1/\tau = v_F \int (2\pi/\hbar)|V_{kk'}|^2 N(E)(1 - \cos\theta)d\Omega/4\pi \qquad (5.5)$$

where $V_{kk'}$ is the appropriate matrix element for the transition from state $\mathbf{k}$ to $\mathbf{k}'$ and $N(E)$ is the density of states at the appropriate energy, here the Fermi energy of the electrons E_F. Usually $V_{kk'}$ is calculated in the Born approximation using a screened pseudopotential to represent the electron–ion interaction (see section 4.6), modified by the structure factor of the liquid $S(K)$ to take account of diffraction effects of the disordered ionic arrangement in the liquid.

We now put equation (5.5) into equation (5.2), with $d\Omega = \sin\theta\,d\theta\,d\phi$ to get:

$$\rho = [3/(ev_F)^2][v_F 2\pi/\hbar] \int_0^\pi \int_0^{2\pi} S(K)|V(K)|^2(1 - \cos\theta)\sin\theta d\theta d\phi \qquad (5.6)$$

If we now choose the polar axis in the current direction, we can integrate over ϕ since there is symmetry about this direction. Finally, since $(1 - \cos\theta) = 2\sin^2(\theta/2)$ and $\sin\theta = 2\sin(\theta/2)\cos(\theta/2)$, we can rewrite the expression in terms of $\sin(\theta/2)$ and then substitute $\sin(\theta/2) = K/2k_F$ (see equation (5.4)). We then find:

$$\rho = (12\pi/e^2 v_F \hbar) \int_0^1 |V_{\mathbf{kk}'}(K)|^2 S(K)(K/2k_F)^3 \, d(K/2k_F) \qquad (5.7)$$

This is the final expression for the resistivity of a disordered metal whose structure factor is $S(K)$ and Fermi radius k_F.

There are a number of approximations in this result.

1 It relies upon the notion that the electrons are free-electron-like.
2 The Born approximation is used.
3 Linear screening is assumed which implies in effect that the ion cores do not overlap each other.
4 Multiple scattering is ignored.
5 We have assumed that elastic scattering is all that matters so that inelastic scattering and the consequent change of energy have been neglected; these can be incorporated into the argument without too much difficulty but their contribution is not usually very significant.

From equation (5.7) we see that the K^3 factor gives a heavy weighting to contributions to the integral at high values of K, that is, near the upper limit of K around $2k_F$. It is therefore very important to know accurately the value of the pseudopotential in this region. We must also be able to determine accurately the value of k_F, which is determined by the number of conduction electrons per unit volume of the metal.

5.3 The temperature dependence of resistivity

For many simple liquid metals the approximations mentioned above are quite adequate and the theory gives a good representation of the conductivity of many simple liquid metals.

The idea of using the structure factor in this way was introduced by Ziman, following in the steps of earlier workers. By using this idea and combining it with the pseudopotential, he revolutionised the subject and thereby suggested natural explanations of many aspects of liquid metals that had previously been baffling.

One example is the temperature dependence of their resistivity. It is found that the temperature coefficient of resistivity, $(1/\rho)d\rho/dT$, measured at constant volume for the liquid just above the melting point is positive for monovalent metals and negative for divalent metals. The Ziman theory offers a convincing explanation of this observation.

When the temperature changes (at constant volume), the structure factor $S(K)$ is the only quantity that changes significantly so that if we know how $S(K)$ changes, the temperature dependence of the resistivity can be found. To understand how the valence of a liquid metal influences its resistivity we must understand something of the shape of the structure factor and of the pseudopotential, and of their position relative to $2k_F$. These are illustrated in Figure 5.1. The essential features are twofold. First, the shape and position of the first peak of the structure factor plotted as a function of Ka (where a is the mean separation of the ions and varies as $\Omega^{1/3}$) is roughly the same (though not in detail) for all the simple metals. This comes about because as we saw earlier the peak in the structure factor occurs at a value of roughly $2\pi/a$. Similarly the general pattern of the pseudopotential plotted in the same way is approximately the same. Second and very important, the value of $2k_F$ varies systematically with the valence n of the metal. Suppose there are N ions in a

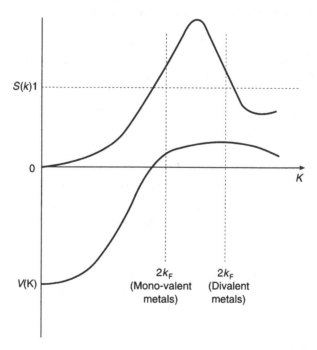

Fig. 5.1 The structure factor $S(K)$ and pseudopotential $V(K)$ of liquid metals (schematic).

volume Ω. In a free-electron-like metal k_F is determined by the relation:

$$4\pi k_F^3 \Omega / 12\pi^3 = Nn \qquad (5.8)$$

which shows that k_F varies as $n^{1/3}$ for a fixed density of ions i.e. for a given value of a. Thus the value of $2k_F a$ is $2^{1/3}$ or about 25 % larger for a divalent metal than for a monovalent metal. From this it transpires that, as indicated in the figure, the value of $2k_F$ falls to the left of the peak in $S(K)$ for the monovalent liquid metals and to the right for divalent or higher-valent metals.

We are now in a position to discuss the temperature dependence of ρ. As the temperature increases the liquid becomes more disordered and so tends to become more gas-like. We can therefore expect that $S(K)$ will tend towards unity at all values of K. Where $S(K)$ is below unity it tends to rise and where it is above unity to fall. This means that at the lower values of K which determine the value of ρ for the monovalent metals (remember that in the integral in equation (5.7) K goes from 0 to $2k_F$) the value of $S(K)$ everywhere increases, thereby increasing the value of the integral and hence that of ρ. So for these metals ρ increases with temperature and the temperature coefficient is positive. For the divalent metals, however, the main contribution to the integral comes from around the region of the peak in $S(K)$ where now the values of $S(K)$, being greater than unity, are decreasing. So ρ diminishes and its temperature coefficient is negative.

This is just one example of the successes of the Ziman theory. It can also account for the magnitude of the resistivities of simple liquid metals at their melting points as illustrated in Table 5.1.

Table 5.1 *Resistivity of some simple liquid metals.*

Metal	$\rho_{exp.}$ ($\mu\Omega$ cm)	ρ_{theory} ($\mu\Omega$ cm)
Li	24.7	25
Na	9.6	7.9
K	13.0	23
Rb	22.5	10
Cs	37	10
Zn	37	37
Al	24	27
Pb	95	64

Data from Faber (1969)

The theory can also tackle the problems of the conductivity of liquid metal alloys. Now instead of the structure factor of the pure metal the partial structure factors of the alloy must be used; in a binary alloy of components A (concentration c_1) and B (concentration c_2), these are S_{AA}, S_{BB} and S_{AB} as we have already discussed in relation to the structure of metallic glasses in section 2.5. The pseudopotentials of the two constituents V_A and V_B are used to describe the scattering potential and the factor $|V_{kk'}|^2 S(K)$ for a single component is replaced by:

$$
\begin{aligned}
V_A^2(K)c_1[S_{AA}(K) - 1] &+ V_B^2(K)c_2[S_{BB} - 1] \\
&+ 2V_A(K)V_B(K)c_1c_2[S_{AB}(K) - 1]
\end{aligned}
\tag{5.9}
$$

The Lindhard screening function is that appropriate to the density of the conduction electrons in the alloy. To calculate this density we need to use the valence of the two components, their concentrations and the molar volume of the alloy at the composition under study. Thereafter the calculation is analogous to that above.

Strangely enough we have been able to discuss the resistivity of the liquid metals, which exist only at high temperatures, in terms of elastic scattering because this is so overwhelmingly important in both the magnitude and the temperature dependence of the resistivity. On the other hand when we come to the metallic glasses, which so closely resemble liquids except that they can exist at low temperatures, we have to take into account inelastic scattering. This means that before we consider glasses we have to familiarise ourselves with some aspects of phonon behaviour in disordered systems, in particular, how they scatter electrons.

6

Phonons in disordered systems

6.1 Normal modes in glasses

Thermal energy causes the ions in a metallic glass, as in a crystal, to vibrate about their mean positions; in a glass there may be additional ionic motion in which ions actually shift between two or more sites but we ignore this for the present. The complex vibrational motion can, as a first approximation, be resolved into a superposition of normal modes, each of which is to this approximation a harmonic motion independent of all the other modes. This ignores anharmonicity and tunnelling modes, which can be very important in glasses. For our present purposes we take the normal mode description as adequate but bear in mind its limitations. These modes introduce into the solid changes in charge density that are periodic in time and cause corresponding changes to the potential seen by the conduction electrons. These changes scatter the electrons.

When such harmonic motions are quantised we associate with each mode phonons in accordance with the intensity of the particular mode. In disordered materials the normal modes of vibration exist although they are not necessarily extended waves; some may be localised to the neighbourhood of particular ions. As long as the vibrations are quasi-harmonic, however, phonons are a valid concept in disordered materials although it may not be possible to assign to them a well-defined wave vector if the mode is strongly localised. Specific heat measurements on metallic glasses have demonstrated that such glasses show larger anharmonic effects at higher temperatures than do crystals (as might be expected); nevertheless we shall treat the phonons in these materials as essentially the same as in crystals, except for their spectrum of frequencies and the general caveat noted above.

55

6.2 Scattering of electrons by phonons

What is the effect of phonons on electrons in metallic glasses? A phonon
of frequency ω and wave vector $\mathbf{q}$ interacting with an electron of energy E
and wave vector $\mathbf{k}$ increases the electron energy by $h\omega$ and changes its
wave vector to:

$$\mathbf{k}' = \mathbf{k} + \mathbf{q} \qquad (6.1)$$

Alternatively the electron can emit such a phonon, reducing the energy of
the electron and changing the electron wave vector as in equation (6.1)
with a minus sign. Because there is an energy exchange, such processes
are termed *inelastic*.

But equation (6.1) represents a rather restricted class of processes. In
addition, the electron may interact with a suitable Fourier component $\mathbf{g}$
of the potential inside the metal as well as with the phonon $\mathbf{q}$. Here we
think of the potential as described by the structure factor and, for example, the screened electron–ion pseudopotential. Previously we have
labelled such Fourier components by $\mathbf{q}$ but since $\mathbf{q}$ is conventionally
used to denote a phonon wave vector, I thought it important to retain
that usage. Moreover the vector $\mathbf{g}$ is sometimes used to denote a vector in
the reciprocal lattice of a crystal and this has some analogy with the
Fourier component here. Interactions of this more general form are possible provided that:

$$\mathbf{k}' = \mathbf{k} + \mathbf{g} + \mathbf{q} \qquad (6.2)$$

To find out how the resultant scattering depends on temperature, we
look at the consequences first at low temperatures and then at high
temperatures. In considering these questions it is important to bear in
mind that as the temperature changes from the absolute zero to room
temperature, the Fermi wave vector k_F scarcely changes in size whereas
the predominant magnitudes of the phonon wave vector q go from zero
to about the same size as k_F. The energy of the electrons at the Fermi
level E_F hardly alters: that of the typical phonons increases from zero to
$k_B\theta_D$ where θ_D is the Debye temperature of the glass, typically around
250–350 K. Even at its highest value the phonon energy is only of the
order of 1% of that of the Fermi electrons. These striking differences
between the electrons and the phonons are important in their mutual
scattering processes.

6.3 Low-temperature scattering of electrons by phonons

First we consider only phonons spontaneously generated in the solid. In addition there are phonons generated, as it were, as a by-product of the scattering of the electrons by the ions (the ions are set in motion by the scattering); their number is determined by the Debye–Waller factor, which we discuss below in section 6.3.2.

By low temperatures here we mean temperatures that are small compared to the Debye temperature θ_D of the glass. The Debye temperature is a measure of the highest frequency ω_{max} in the normal mode spectrum, defined so that $\hbar\omega_{max} = k_B\theta_D$. If therefore we consider temperatures less than, say, one-twentieth of θ_D, the phonons predominantly excited at these temperatures will have frequencies of at most about $\omega_{max}/20$. Moreover the wavelength will be at least twenty times bigger than at ω_{max}, which is itself about the interionic spacing a. At low temperatures therefore we are dealing with long-wavelength phonons with correspondingly small q vectors, less than about $\pi/10a$. We must also remember that in all metals with typically a few conduction electrons per ion, k_F is about $2\pi/a$ so that the value of q for phonons at low temperatures is always small compared to the wave vectors of the Fermi electrons. As a first approximation therefore we can neglect q in equation (6.2); we then see that the predominant g-vectors will be almost the same as those that promote strong elastic scattering and mainly determine the resistivity of the glass. Thus this type of low-temperature inelastic scattering will be in proportion to the total resistivity of the glass at low temperatures.

To see how the scattering depends on temperature, we note that the final state $\mathbf{k}'$ must satisfy equation (6.2) and its energy E' must differ by $\hbar\omega$ from that of $\mathbf{k}$. If k, k' and q are well defined, the tip of $\mathbf{q}$ must therefore lie on a two-dimensional surface in k-space with the consequence that the number of phonons at temperature T that can participate in the scattering is the number whose wave vectors measured from a common origin terminate in such two-dimensional surfaces. In the temperature range involved here, where the isotropic continuum Debye model should be satisfactory, this number varies as T^2. (The total number in three dimensions varies as T^3 as seen in the Debye T^3 law.)

As we saw above, the $\mathbf{g}$-vectors involved will be predominantly those with large-amplitude Fourier components in the structure factor and in the pseudopotential. These typically occur at large scattering angles, which therefore make the final scattering angles effectively independent of the tiny values of q involved at low temperatures. Thus the $(1 - \cos\theta)$

factor that we discussed earlier does not vary with temperature. So the only temperature-dependent term is that due to the number of participating phonons and consequently the inelastic scattering due to phonons in a disordered metal or alloy at low temperatures increases as T^2.

As we shall soon see, however, not all low-temperature phonons are fully effective in scattering electrons when their wavelength is greater than the electron mean free path. We may also suspect that, with the high degree of disorder characteristic of glasses, k, k' and especially q will not be well defined and processes other than those allowed by equation (6.2) will occur. Indeed, more detailed studies suggest that the temperature variation is different for longitudinal and transverse phonons and together the temperature variation, although predominantly as T^2, can include powers of T from 2 to 4.

6.3.1 Phonon drag

The T^2 temperature dependence of the scattering that we have looked at is very different from that of, for example, the pure crystalline alkali metals where the resistance at low temperatures increases as $\exp(-c/T)$, where c is constant.

One reason for the big difference in this case is that in the alkali metals the phonon distribution is greatly altered by the electron flow. We can say that the electron current tends to drag the phonon gas along with it and, unless there is a mechanism that tries to hold the phonon distribution at rest with respect to the lattice, the phonons can drift with the electron current and, in extreme cases, almost cease to impede its flow. At low temperatures in the alkalis this is indeed what happens. In metallic glasses, however, the disorder in the ionic positions is enough at most temperatures, except perhaps below about 1 K, to scatter the phonons and keep them from drifting. This completely alters the response of the electrons to the phonon scattering and makes possible the T^2 dependence described above.

The general absence of phonon drag in disordered systems is also important in the thermopower of metallic glasses.

6.3.2 The Debye–Waller factor

This is not yet the whole story. In addition to the phonons that are spontaneously generated by the thermal motion of the ions, there are other phonons generated by the scattering of the electron from the disorder of the ions; these are also inelastic processes. When an electron is scattered by the ionic disorder at low temperatures such scattering is usually elastic, with no phonon generated. However, as the temperature is raised from absolute zero the probability of inelastic scattering from the disordered ions increases and the probability of elastic scattering correspondingly decreases. Thus the elastic scattering contribution to the resistivity (usually called the residual resistivity) decreases as the temperature rises, this decrease varying as T^2 and in direct proportion to the residual resistivity ρ_0. It is accounted for by the so-called Debye–Waller factor, which tells us the fraction of events at any temperature that are inelastic.

This decrease of resistance is, at low temperatures, more than offset by the increased inelastic scattering, which causes the resistance to increase, also as T^2. Moreover this resistance is about twice as big as the reduction caused by the fall-off in elastic scattering. (At high temperatures the elastic scattering falls off in direct proportion to the temperature as does the associated inelastic scattering but now the two terms are equal and opposite and so cancel.)

The Debye–Waller factor describes an effect which is of wider significance than is apparent from the way I have introduced it. It was calculated originally, by the scientists whose name it bears, to account for the temperature dependence of X-ray scattering in crystals. If we consider a crystal at absolute zero and allow its temperature to rise we find that the intensity of a Bragg reflection falls and correspondingly scattering at neighbouring frequencies associated with the absorption or emission of phonons increases. Likewise in the Mössbauer effect the recoilless emission and absorption of γ-rays from nuclei in a crystal has its greatest intensity at absolute zero and diminishes as the temperature rises. The importance of all this is that there are two quite distinct classes of process involving the lattice. In one, the lattice responds to the stimulus (scattering of X-rays or conduction electrons or the emission of a γ-ray by a nucleus) with absolutely no change in its degree of excitation: such processes are truly elastic as far as the lattice is concerned. In the other, phonons are emitted or absorbed and so the processes are inelastic. Even in the liquid phase, surprisingly enough, such completely phononless processes, though rarer, still occur. Of course although I have referred to *crystals* in this discussion, it applies equally to glasses.

Later on we shall be greatly concerned with scattering processes that leave the wavefunction of the electron coherent with its state before scattering. These elastic processes that we are discussing precisely satisfy that condition whereas all inelastic processes (in which the quantum energy state of the scatterer is changed) introduce incoherence, i.e. the phase of the electron wavefunction after scattering is randomised.

The existence of these phononless processes so dramatically demonstrated by the Mössbauer effect seems contrary to one's intuition and one therefore immediately suspects that quantum effects are at work. This is, however, not so. A classical analogue is to be found in frequency modulation. In classical terms we can think of the γ-ray or X-ray or electron wave as being frequency-modulated by the vibrations of the ion or ions in the solid. In such modulation, however, there is always a contribution corresponding to the unmodulated carrier. Here this is equivalent in quantum terms to the phononless processes.

The Debye–Waller factor depends only on the vibrational properties of the scattering system and is similar for all stimuli. What causes the scattering depends of course on the nature of the incident wave: for X-rays it is the electron density; for neutrons the nuclear scattering cross-section or scattering length; and for electrons the electrical potential. In all cases the scattering is modified by the motion of the ions.

There is a further interesting feature of the Debye–Waller factor: even at the absolute zero the elastic scattering is modified because the zero-point motion of the ions makes their positions 'fuzzy' and effectively alters their form factor. Thus the zero-point motion, unlike zero-point energy, is a physical reality whose effects show up in, for example, this change of form factor. This zero-point motion does not upset the coherence of the electron waves that propagate through it and produces no electrical resistance.

6.3.3 Phonon ineffectiveness

There is a further effect that has been widely discussed in relation to metallic glasses but has only recently been clearly identified, not in a glass but in concentrated potassium–rubidium alloys. It involves the Pippard ineffectiveness condition (strictly its inverse), according to which a phonon becomes ineffective in scattering an electron when the mean free path of the electron l is very short compared to the wavelength λ of the phonon. In classical terms it is as if the electron begins to interact with a sound wave but is scattered by an ion after seeing only a tiny fraction of the sound wave (the phonon) and before it has had chance to recognise the existence of the wave motion. This means that as we go to low temperatures and as the dominant phonon wavelength gets longer and longer, the scattering of the electrons as discussed above falls below that predicted and is ultimately killed. It is as if phonons for which $q < \pi/l$ are incapable of scattering electrons. If at high temperatures there are enough phonons that have short wavelengths the effect is masked but shows up as the temperature falls. This is found experimentally and with quantitative agreement in K–Rb alloys at temperatures below 1 K.

This effect is thought to have been seen in metallic glasses but the disorder of the glassy state means that the criterion in terms of a single wave vector q is hard to apply since the phonons may not be representable by a small range of wave vectors clustered around $\mathbf{q}$.

It will, I think, be clear from what we discussed earlier that the Debye–Waller factor is not influenced by this sort of effect.

6.4 Scattering of electrons by phonons at high temperatures

We return now to the scattering of conduction electrons by spontaneously generated phonons, this time at high temperatures, i.e. at temperatures comparable to the Debye temperature.

At these temperatures we can use the Einstein model to describe the ionic motion in a glass. In this, each ion is assumed to vibrate independently and harmonically about its mean position in the average field of all its neighbours. Incidentally this, if taken literally, would be the ultimate in localisation of a normal mode. The position of the mode is essentially represented as a δ-function at the site of the ion and so would require, as we know from the Fourier transform of a δ-function, an infinite range of q-vectors to represent it. This is an extreme and rather unphysical example of what I referred to earlier in discussing local modes. Nonetheless we can still use the model at high temperatures provided we recognise its limitations.

Consider now the effect of a single ion displaced by a small amount X in the x-direction. If the potential at a point (x, y, z) due to the ion is $V(x, y, z)$ when in its mean position, the potential at the same point when the ion is displaced is $V(x - X, y, z)$ on the assumption that the potential moves bodily with the ion. This neglects any motion of the conduction electrons to screen out the effect. The change in potential is thus:

$$\Delta V = V(x - X, y, z) - V(x, y, z) = -X(\partial V/\partial x) \qquad (6.3)$$

The matrix element associated with scattering an electron from states $\mathbf{k}$ to $\mathbf{k}'$ is therefore:

$$V_{\mathbf{k}\mathbf{k}'} = -X \int \psi_{\mathbf{k}'}^{*} (\partial V/\partial x)\psi_{\mathbf{k}} \, d\tau \qquad (6.4)$$

where the integral is over one ionic cell only, in order to take into account in a crude way the screening effects to which reference has already been made. To calculate the transition probability of scattering the electron from states $\mathbf{k}$ to $\mathbf{k}'$ we need the square of this matrix element. This is proportional to X^2 or on average to the mean value of X^2. Likewise for displacements in the other two independent coordinate directions. The upshot is that the probability of scattering from this ion is proportional to its mean square displacement, which at high temperatures is proportional to the absolute temperature T. We therefore conclude that the scattering of electrons at high temperatures is also proportional to T. In many pure metals the behaviour of the resistance shows that this is indeed true. In concentrated alloys and glasses the resistance itself does not directly reflect this temperature dependence; the effects are more subtle, as we shall see.

7

Interactions and quasi-particles

7.1 The validity of the independent electron picture

We have treated the electrons as effectively independent particles subject to occasional scattering processes even though we know that there are strong Coulomb forces between electrons and between electrons and ions. This picture certainly has some validity which can be partly understood in the following way. First of all, as we have seen in Chapter 4, the range of the Coulomb interaction between electrons is screened out over a distance of the order of the interionic separation because the conduction electrons are attracted to the neighbourhood of the positive ions and so produce electrical neutrality when viewed from a short distance away. Thus the cross-section for scattering of an electron is of the same order as that of an ion, i.e. of atomic dimensions.

Second, the Pauli exclusion principle drastically reduces the number of processes by which conduction electrons can interact and be scattered by other conduction electrons. We can see this from the following argument. Consider an electron gas at absolute zero with all the states up to E_0 filled and those above empty. Assume that we give one electron a small amount of energy ϵ above E_0. It can only be scattered by another electron in the Fermi sea if, after the collision, both particles have empty states of the right energy to go to. This means that, since energy is conserved in the collision, the initial state of the second electron must lie within an energy range ϵ of the Fermi level; otherwise the collision could not raise its energy above E_0 where there are empty states. Thus the number of electrons that can take part in scattering is severely reduced, being proportional to the number of states from which the second electron can be chosen. If the density of states at the Fermi level is $D(E_0)$, the number of states in an energy range ϵ is $D(E_0)\epsilon$.

In a classical scattering process the scattered particle can go into any energy state allowed by conservation of energy. In the Fermi–Dirac gas, however, both particles after colliding must find empty states i.e. states above E_0. So the range of energy available to the scattered particle is limited to between E_0 and $E_0 + \epsilon$. This therefore further reduces the possible scattering processes and the probability of scattering is proportional to a further factor of $D(E_0)\epsilon$. This shows that the probability of scattering of one electron by another in a given metal varies as ϵ^2. In particular this means that an electron at the Fermi level and at absolute zero (for which therefore $\epsilon = 0$) cannot be scattered in this way.

We can apply this argument to find the temperature dependence of electron–electron scattering. Suppose the electron gas is at temperature T, which we take to be small compared to the Fermi temperature T_F of the electrons ($k_B T_F = E_0$). The fraction of electrons that are excited is of order T/T_F. Thus by an argument similar to that just outlined we find that the effect of electron–electron scattering is reduced by a factor of $(T/T_F)^2$ compared to that to be expected classically. At room temperature this is typically (see Table 3.1, p. 24) of order 10^{-4}. Moreover since the cross-section for an electron in electron–electron scattering is similar to the cross-section of an ion in electron–ion scattering, the resistivity due to electron–electron scattering is small compared to normal high-temperature resistivities. Only at low temperatures where the scattering by phonons has substantially diminished is electron–electron scattering manifest as a T^2 term in the resistivity.

7.2 Quasi-particles

Landau tackled the problem of interacting particles from a different point of view, originally to try to understand the properties of liquid He^3. His starting point is not the independent particle picture but a strongly interacting system of N particles (the Fermi liquid) with a small number of excitations of low energy ϵ, very much smaller than the Fermi energy E_F. Landau called these excitations 'quasi-particles' or, here, 'quasi-electrons' because they are the interacting analogues of the non-interacting particles. If we start with N non-interacting electrons and allow the interaction between them to increase gradually, the states of the non-interacting electrons evolve into excitations of the strongly interacting Fermi liquid, with which they have a one-to-one correspondence (at least in the low-lying excitations). The quasi-electrons have charge e and spin $1/2$ and so obey the Pauli exclusion principle. The argument given in the previous paragraph about its influence in limiting the effects of the interaction still apply and a quasi-electron of energy ϵ above the Fermi level has a probability of being scattered that varies as ϵ^2. In particular it vanishes when $\epsilon = 0$. Therefore at the Fermi level, these quasi-particles have the same properties as their independent counterparts and, because of the restriction on the scattering, they remain, at low energies, weakly interacting entities.

As their energy moves away from E_F, the interactions cause their properties to differ from those of independent particles. The quasi-particles are still labelled by their k-vectors but now, because of the interaction, their energy $\epsilon(k)$ is altered and the excitations have a limited lifetime. The state can be represented by a complex number whose real part is the new energy ϵ (the self-energy) and the imaginary part gives the lifetime γ of the quasi-particle. Quasi-particles are well defined if the uncertainty in their energy h/γ is small compared to ϵ, measured from the Fermi level, and under these conditions we can treat the quasi-electron as a nearly free particle, although with a different effective mass. For this reason the density of electron states is altered; nevertheless it changes smoothly with energy through the Fermi level.

The quasi-particle concept explains and justifies the treatment of the electrons in a metal as weakly interacting particles and we shall use this idea throughout the rest of the book, although its limitations have to be borne in mind, especially when we treat the enhanced electron interaction brought about by disorder.

Another important aspect of the quasi-particle is its interaction with the ions, which are set in motion as an electron passes near. The quasi-particle can absorb this kind of interaction into its properties (for example, its effective mass). As we shall see this ion-mediated interaction turns out to be of great importance not only in superconductivity, but also in electron–electron scattering and in the thermopower.

An electron in a metal repels other electrons that approach it; it has what is often called a 'correlation hole' around it. This hole can also be thought of as arising from displaced electrons and its effects reduce the interaction with other electrons, thereby helping to explain why the conduction electrons can be treated as almost free particles unless they are specifically scattered by, for example, impurities or phonons.

7.3 Particle–hole pairs

When we are dealing with a highly degenerate gas of electrons as in a metal, there are, at normal temperatures and particularly at low temperatures, comparatively few electrons excited above the Fermi level. This also means that there are correspondingly few empty states in the Fermi sea of the remaining electrons. These ideas have been formalised and the concepts and terminology that result are often used. Let us briefly look at them here since they will be of use later.

The idea is to simplify the description of our Fermi system by recording just changes from the ground state and so referring only to the electrons excited above the Fermi level and the unoccupied states below it. The rest of the electrons (or Fermions) are omitted as illustrated in Figure 7.1. The word 'particle' then takes on the special meaning of an electron above the Fermi level. The empty states are called 'holes' in this particle–hole description and, with the rest of the Fermi sea removed, behave like anti-particles or, in some respects, like the holes familiar in the band theory of metals and semiconductors. They have states below the Fermi level.

The hole represents a particle removed from the system so its energy is negative. The wavefunction of the hole has the same form as that of the absent particle but suitably modified. For example if we consider a particle of energy E_1 whose wavefunction has the time-dependent part $\exp(-iE_1 t/\hbar)$, the wavefunction of the hole in the same state (with energy $-E_1$) would be $\exp(+iE_1 t/\hbar)$. This is then reinterpreted by associating the

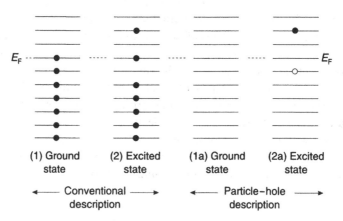

Fig. 7.1 Particle–hole description of degenerate Fermi system. It shows only the excited electrons and the vacant states below the Fermi level.

change of sign, not with the energy, but with the time; in this way the hole becomes a particle moving backwards in time. This convention was introduced by Feynman in his work on positrons and is used in interpreting Feynman diagrams.

Further properties of a hole are as follows:

the hole charge $= -$ the particle charge;

the hole spin $=$ the opposite of the particle spin;

the hole wave vector $\mathbf{k} = -\mathbf{k}$ of the particle.

This provides a formal framework which is convenient for calculations that involve, for example, the electrons in a metal or alloy and can be applied to quasi-particles. It is a formalism frequently used in the literature and some of it spills over into common use.

7.4 Electron–electron interaction mediated by phonons

We know that the electron–phonon interaction causes the scattering of conduction electrons by phonons. It also shows itself in the equilibrium state of metals as the commonest cause of superconductivity. The theory of superconductivity devised by Bardeen, Cooper and Schrieffer (BCS) is based on this interaction; it gives rise to an attraction between electrons near the Fermi level and thereby to the superconducting state. Even in those metals that do not show superconductivity, such as the alkali metals, the interaction shows up as a modification of electron–electron scattering and also as an enhancement of the electronic heat capacity.

Let us now look at this electron–phonon interaction in classical terms to see how it brings about the superconducting state. As an electron moves through the positively charged ions they are drawn to it by the Coulomb attraction and are set into oscillatory motion. These create a periodic variation of positive charge which influences the energy of other electrons. If the motion of another electron is suitable in timing and position in relation to this oscillating charge it can use the lower potential of the excess positive charge to reduce its own energy. The resulting interaction between the two electrons of opposite momenta and spin (referred to commonly as a Cooper pair) is thus attractive. When many electrons cooperate to produce a dynamic pattern that exploits this reduction in energy, it can be sufficient to offset the lowering of their entropy S which necessarily comes about through sustaining the ordered pattern.

The stable state is that of lowest free energy $E - TS$. We assume that the major changes in E and S are due to the electrons, i.e. that the ions are essentially unaffected. Thus since the penalty from the entropy term TS gets smaller as we lower the temperature the cooperatively ordered state of lower energy eventually, at some specific temperature, becomes the stable state. This state is the superconducting state.

We can put the matter in another way. Suppose the metal is at the absolute zero in its ordered superconducting state and we allow its temperature to rise a little. There are now unpaired electrons thermally excited that rush about in a completely uncoordinated way, each one a loose cannon, as it were, seeking to play havoc with the disciplined ranks of the superconducting electrons. As the temperature rises the effect of this disorder grows. Eventually when the temperature is high enough, the disorder is sufficient to disrupt the coherence of the super-conducting electrons and the metal goes over to the normal state.

We are interested in the consequences of the electron–phonon inter-action for the *normal* state of a metal. In order to understand some of the detail we have to recognise that the ionic vibrations are quantised so that their interactions with the electrons take place through phonons, although in this instance these are not the thermally excited phonons that cause electrical resistivity. The interaction here comes about through 'virtual' processes, which can exist even at absolute zero. They consist of the excitation and de-excitation of phonons of energy ΔE (say) in a time interval Δt so short that $\Delta t \cdot \Delta E < h$. An electron involved in the interaction by the absorption of a virtual phonon must have an unoccupied state into which it can go. Thus the only electrons involved in the interaction are those at the top of the Fermi distribution. If ω_{max} is the highest frequency of oscillation sustainable by the ions, only electrons within an energy range of $\hbar\omega_{max}$ of the Fermi level can participate in the interaction. If for convenience we use a Debye model for the ionic vibrations, we can use the approximation that $\hbar\omega_{max} = k_B\theta_D$.

From this discussion it will, I think, be evident that as the temperature is raised, thermal fluctuations will tend to disrupt this interaction. Empty states that would otherwise be available for the virtual processes are occupied by thermally excited electrons and so weaken the interaction; in fact when temperatures of the order of the Debye temperature are reached, the interaction is completely destroyed. The thermal excitations of the electrons are now so fast and violent that the lattice vibrations can no longer follow them.

Our picture therefore is of a Fermi electron moving through the ions and setting them in motion. This alters the energy levels to which the electron has access and affects electron states that lie within $\pm k_B \theta_D$ of the Fermi level. Since moving an electron entails also setting in motion the other electrons or ions with which it interacts, one can think of the electron as having a greater effective mass so that (see equation (3.12)) the density of states within $\pm k_B \theta_D$ of the Fermi level is increased and the electron velocity correspondingly decreased. The electrical conductivity is thus unaffected but the result can be seen in the low-temperature electronic heat capacity. For example in the crystalline alkali metals the measured values are significantly bigger than the theoretical values calculated from the band structure of the metals. When this 'band' contribution to the heat capacity is corrected by the calculated enhancement due to the electron–phonon interaction (and a smaller correction due to electron–electron interaction) agreement is achieved. The effect is also to be expected in metallic glasses but it cannot be inferred from the measured electronic heat capacity because we cannot accurately establish what the unenhanced value should be; the band structure is not known accurately enough.

As we have seen the interaction also leads to an attractive force between electrons that can overcome the Coulomb repulsion and lead to superconductivity. In the normal state the attraction modifies the electron–electron scattering and its associated resistivity. For a long time such scattering was attributed solely to the Coulomb repulsion until finally it was realised that the phonon-mediated attraction is also involved. This has the additional feature that, unlike the Coulomb repulsion, it is temperature-dependent: its strength is greatest at low temperatures, weakens as the temperature rises and finally vanishes completely. Finally this interaction plays an important role in the thermopower of metallic glasses as we shall see in Chapter 15.

The strength of the electron–phonon interaction can be characterised in various ways. The electron–phonon mass-enhancement factor λ_{ep} (which determines the enhancement of the electronic heat capacity at the lowest temperatures) is defined as:

$$\lambda_{ep} = \int [\alpha^2 F(\omega)/\omega] d\omega \qquad (7.1)$$

where α is the electron–phonon coupling constant, ω is the phonon frequency and the combination $\alpha^2 F(\omega)$ is the Eliashberg function, important in superconductivity. This function also appears in the probability of scattering of electrons at the Fermi level by phonons:

$$1/\tau_{ph} = \int [\alpha^2 F(\omega)/\sinh(\hbar\omega/k_B T)]d\omega \qquad (7.2)$$

Here τ_{ph} is the lifetime of the electron state, not the transport relaxation time. In crystalline metals, α is normally constant at small values of ω while $F(\omega)$ varies as ω^2 so that the probability of scattering at these frequencies (i.e. at low temperatures) varies as T^3. On the other hand, calculations on weakly scattering crystalline alloys and metallic glasses show that the function varies as ω at low frequencies so that the lifetimes now vary as T^2 at low temperatures, a feature derived by other means above.

7.5 Collective electron modes

We have seen how the concept of a quasi-particle can restore the idea of independent electrons in metals while still taking account of the mutual Coulomb repulsion and the phonon-mediated attraction between electrons. Nonetheless, one is tempted to ask: Why do we not use collective modes to describe the highly interacting electron–ion assembly? After all, the motion of the ions is described, not in terms of the individual ions, but through the collective modes, the normal modes of vibration of the assembly and their quantisation as phonons.

Such a description of the interacting ions and electrons in terms of their collective modes was carried out by Bohm and Pines. The collective modes that are familiar in such an assembly are the plasma modes that involve the positive and negative ions oscillating in antiphase.These are not usually excited. There are other modes which together describe the screening of the electrons and lead to results not too unlike the exponential screening derived by more elementary and less rigorous methods (e.g. the Fermi–Thomas approximation). Both descriptions, that in terms of quasi-particles and that in terms of collective modes, have their appropriate place and usefulness in the theory of metals.

8

Transition metals and alloys

The obvious success of the Ziman theory does not extend to the liquid transition metals and, as we shall see in the next chapter, the Hall coefficient of a number of glasses containing a substantial proportion of transition metal is positive, thereby posing a powerful challenge to conventional theories. Before we try to compare theory and experiment, however, let us look at some of the important properties of transition metals and their alloys, in both crystalline and glassy forms.

8.1 Crystalline transition metals

A transition metal is one whose atoms have incomplete d-shells, such as iron or tungsten. Typically in the free atom there are also s electrons from a higher electron shell, for example, there may be two 3d-electrons and one 4s. In the solid state the wavefunctions overlap and the single states of the free atom spread out into bands whose electrons can therefore take part in the conduction process. The s-levels broaden much more than the d-levels as the atoms get closer. This is because the s-electrons come from the outer reaches of the atom with wavefunctions that overlap strongly with those of their neighbours in the solid. The d-electrons by contrast are more tightly bound within the atom and so in the solid form a much narrower band whose electrons tend to have much lower velocities. The d-electrons, because they are so much more confined, tend to retain more of their atomic character in the solid state including some of the angular momentum associated with their quantum number $l = 2$.

In general the s- and d-states in the free atom from which the corresponding bands arise are close in energy and so the broadening causes the d-band and the s-band to overlap in energy as illustrated in Figure 8.1. There the d-band is seen as a rather craggy structure straddling the Fermi

71

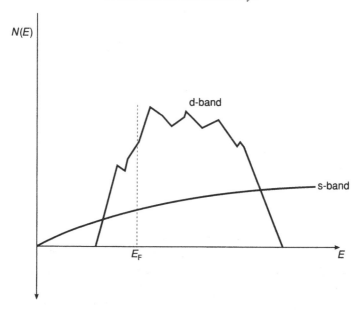

Fig. 8.1 The density of states of overlapping s- and d-bands (schematic).

level, which lies across both s- and d-bands indicating that neither is full. Where precisely the Fermi level lies alters systematically with the position of the element in the periodic table: it lies at low energies for elements with few d-electrons and rises progressively through the band as the number of d-electrons increases. An important feature of the d-band is its high density of states, a consequence of its narrowness in energy and the fact that it must contain ten electron states per atom.

Cu, Ag and Au are sometimes referred to as late transition metals although the d-band in the pure metals is completely filled. Nonetheless the electrons on some parts of the Fermi surface of these metals do have quite strong d-character. What does it mean to talk about s-, p- or d-character? The phase of the electron wavefunction changes in a free-electron-like manner as it propagates through the lattice but inside the ion cores the wavefunction takes on something of the character of an electron in the field of the corresponding nucleus. Thus an s-like electron partakes of an s-type atomic wavefunction in the neighbourhood of the nucleus and so has a non-zero probability of being found there. This shows up in experiments such as the measurement of the Knight shift in nuclear magnetic resonance (NMR) and contrasts with the behaviour of p- or d-like wavefunctions which have almost vanishing probability of being found at the nucleus.

The character of the electron wavefunction may also be important in scattering: suppose that in a scattering centre such as an impurity all the scattering potential is concentrated in the region around its nucleus. Such an impurity will tend to scatter s-like electrons strongly because they have a large amplitude there whereas the p- or d-like electrons will be largely unaffected.

A further characteristic of electrons as they propagate through the ion cores is that they can locally participate in the angular motion of the corresponding atomic state. When the atom is in a solid the electron states may be seriously modified through interaction with the electric field of its neighbours. Nonetheless the spin–orbit coupling that acts on electrons in the free atom also affects conduction electrons and has important consequences for transport properties, in particular the Hall coefficient and quantum interference effects. In fact, of course, s-like wavefunctions have no angular momentum about the nucleus and do not participate in the spin–orbit coupling whereas the p- and d-like wavefunctions do.

Figure 8.2 illustrates the overlap of the d- and s-bands by showing the energy E of the electron versus its wave number k. Where the E–k curves for s- and d-electrons cross a phenomenon known as hybridisation may occur. If the symmetry of the wavefunctions of the s- and d-electrons in the particular crystallographic direction considered is appropriate the two curves as it were repel each other and do not cross (Figure 8.3). The analogy here is with two independent classical oscillators of the same frequency. If now the two are coupled to each other by a harmonic coupling, two normal modes (harmonic oscillations of a single frequency) result, one of a lower and one of a higher frequency than the original. The lower and higher electron energy states, in the neighbourhood of the place where they would have coincided if there was no interaction, originate in this way. The character of the wavefunctions in this region is changed so that the d-electrons are no longer purely d-like but assume some s-character and the s-electrons change in the opposite sense. So each becomes a hybrid with the other, just as each classical normal mode involves both oscillators. In the region of hybridisation the values of k for a given E are not necessarily greatly changed but the corresponding velocities (measured by the slope of the curves) can alter dramatically. This feature is of great importance in some metallic glasses.

Because the s-electrons usually have higher velocities than d-electrons, we tend to think of them as the main current carriers but if the Fermi

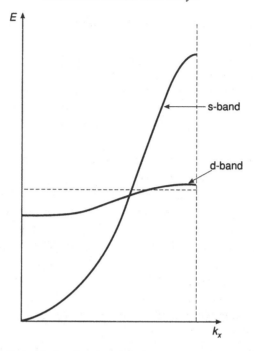

Fig. 8.2 The E–k relation for s- and d-electrons without hybridisation.

level lies in the region of hybridisation the distinction between the two sets is blurred and their velocities, as we have seen, may be substantially changed.

8.2 The Mott model

We now consider the Mott model of crystalline transition metals and how far it applies to amorphous metals and alloys. The main feature of the Mott model is that in a transition metal or alloy there are two rather separate groups of mobile electrons, the s-like and the d-like. The electric current is assumed to be carried mainly by the more mobile s-electrons but these can be scattered by phonons into empty d-states at the Fermi level. Since these d-states have a high density in energy compared to an sp-type metal the resistivity due to phonon scattering is greatly enhanced in the transition metal compared to its non-transition metal counterpart. This is indeed broadly speaking true.

On the other hand, one has to be more careful in the analysis of the residual resistivity of an alloy that contains a transition metal (call it T)

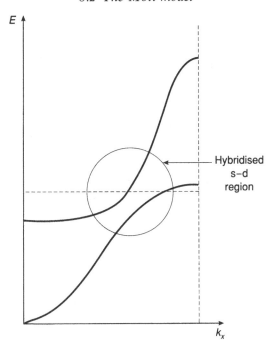

Fig. 8.3 The *E–k* relation for s- and d-electrons with hybridisation.

mixed randomly with a non-transition metal N, say. N is a metal with no d-states or one in which all the d-states are fully unoccupied and lie well below the Fermi level of the alloy.

To see what is involved let us consider what happens in a crystalline alloy. Suppose that the N atoms (e.g. Ag atoms) in the alloy ionise to contribute one s-electron to the s-band but no d-electrons to the d-band since the d-states all lie well below the Fermi level. We assume that T atoms contribute one s-electron to the s-band and one d-electron to the d-band (the precise numbers we choose do not alter the argument except that it is important that the transition metal contributes, as it usually will, to both the s- and d-bands). There are thus occupied s-states on every site, N or T, and so in the alloy we assume that the s-electron wavefunctions overlap and form an s-band over all sites throughout the crystal.

For the d-sites the situation is different because the d-states are more localised and in any case do not exist on the N ions. Thus the d-band, at those concentrations at which it exists, is confined to the T sites. This has important consequences for the transport properties. First of all it means that at sufficiently low concentrations of the transition metal there will be no d-band, just a random arrangement of isolated T ions. These have unoccupied d-states at the Fermi level which mix with the continuum of s-bound states of similar energy to form what are commonly referred to as virtual bound states. These are not true bound states but we may think of the conduction electrons visiting the site and for a short time taking on the character of a d-electron before rejoining the conduction band. These virtual bound states thus scatter conduction electrons but are not numerous enough to form a d-band. Only when the concentration of T is sufficient to provide continuous chains of neighbouring T sites throughout the crystal will a d-band form. This is a so-called percolation problem and we might expect a d-band to form at somewhere around 15% of T.

At the other end of the concentration range, where T is the predominant element, there undoubtedly exists a d-band with its associated high density of states. We might therefore expect strong scattering of the mobile s-electrons from the N impurity ions into these high density d-states. This does not happen for the following reason. The matrix elements for scattering of an s-electron into an s- or a d-state at an N site are given respectively by:

$$\int \psi_s^* \Delta V \psi_k \, dr^3 \quad \text{and} \quad \int \psi_d^* \Delta V \psi_k \, dr^3$$

where ψ_k refers to the incoming s-electron and the other wavefunctions refer to final state s- or d-states, respectively. ΔV is the scattering potential at the site. Since we assume that there is everywhere a typical s-band, ψ_s has a typical amplitude at this site. On the other hand there is, we assume, very little overlap of a d-wavefunction onto an N site from a neighbouring T site. So ψ_d has a very small amplitude and scattering into d-states from the N site is improbable. Thus the high density of d-states is inaccessible and the scattering by an N impurity in a largely T matrix is not unusually large by comparison with that of N in a non-transition matrix. The validity of this argument is illustrated in the dilute alloys at both ends of the crystalline Ag–Pd system.

This argument does not persist, however, at appreciable concentrations of the T element for two reasons. First the scattering at a T site (which is measured by its deviation from the average potential, averaged, that is, over both N and T sites) involves more and more d-character as the concentration of T increases. Second, there may well be more and more hybridisation of the wavefunctions, which therefore are no longer purely s-like. This reason would be specially cogent in a metallic glass.

8.3 Transition metal glasses

Let us now look at some features of the electronic structure of metallic glasses containing transition metals, to compare with what has just been said about the crystalline metals.

The electronic structure of glassy alloys cannot be examined by many of the refined techniques that have been developed for pure crystalline metals because these techniques, for example, the de Haas–van Alphen effect, the radio-frequency size effect, cyclotron resonance and high field magneto-resistance, exploit the long mean free path of the conduction electrons. Positron annihilation and Compton scattering do not rely on this feature and have been used to give information about the momentum distribution of the conduction electrons in glassy materials. They confirm the free-electron behaviour of many of the simple metal glasses. Where d-electrons are involved, however, the results are difficult to interpret.

A successful method that gives information on the electron density of states is that of photoemission, essentially an application of the photoelectric effect. In this technique, high energy monochromatic photons, usually derived from synchrotron radiation, fall on the alloy and the energy spectrum of the electrons emitted gives a measure of their density of states in the glass. To obtain the energy of the conduction band state from which the electron is emitted, you subtract the measured kinetic energy of the emitted electron from the known energy of the photon; the remainder is then corrected for the work function of the glass (the energy difference between an electron at the Fermi level and one at rest just outside the material) to obtain the required energy relative to the Fermi level. It is then assumed that the number of such electrons is proportional to the density of states at that energy. This presupposes that all the electrons in the conduction band interact equally with the photons, which is not true since the probability depends on the wavefunction of the electron involved. Sometimes a correction can be applied but often this is not known or is ignored. This difference in response can, however, be exploited to distinguish the contributions of different types of ion. Figure 8.4 shows the spectra of amorphous $Cu_{60}Zr_{40}$ for photons at two very different energies, 60 eV and 120 eV. The scattering cross-section of Cu d-electrons is much higher than that of Zr d-electrons at photon energies above 70 eV and this makes it possible to identify the Cu d-band as responsible for the lower peak at around -3.5 eV. The other peak seen at higher energy in the first diagram is due to the Zr d-band. It

Transition metals and alloys

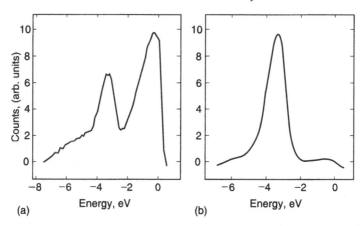

Fig. 8.4 Photoemission from Cu–Zr glass with photons of two different energies: (a) 60 eV (b) 120 eV. (After Greig *et al.* 1988.)

seems likely that the technique can be extended to give more detailed information about the different distributions of s- and d-states.

Many photoemission experiments have been made on metallic glasses because the information they give is so hard to get by other means. An instructive comparison between glass and crystal is made in Figure 8.5, which shows photoemission data on pure Cu and Cu_3Zr_2 in the crystalline form compared to $Cu_{60}Zr_{40}$ in the glassy state. In the Cu results you can see the s-band from the small dip at the Fermi level: in the alloys this is masked by the high density of d-states. The alloy results are surprisingly similar to those of the crystal, although they do show that the crystalline material has sharper features, as one would expect. The peak at lower energies comes from the Cu d-band, which is fully occupied, and the upper one from the Zr d-band, which is incompletely filled, having about two d-electrons per Zr atom out of a possible ten for a full band. There are thus two d-bands associated with the sites of the two types of ion; this is because the bands are well separated in energy. If they were close in energy they would form a common d-band in which d-electrons could move from one site to another having a different ion. We assume that this is what happens to the s-bands, which of course overlap much more both in energy and space.

The data also show that there is some interaction between the two d-bands which shifts their position relative to the pure metal. The Cu d-band is pushed to lower energies in the alloys. This is a general phenomenon whereby the d-band of the late transition metal (here Zr) is pushed

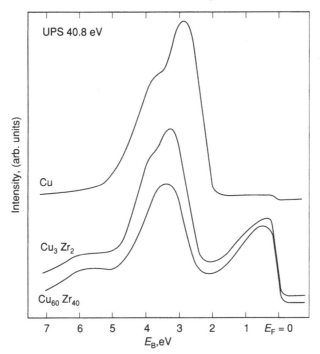

Fig. 8.5 Photoemission from crystalline Cu and Cu_3Zr_2 and $Cu_{60}Zr_{40}$ glass. (After Güntherodt *et al.* 1980.)

to higher energy and the early transition metal (here Cu) is pushed to lower energy.

Photoemission experiments on a series of glassy Zr alloys have shown that in alloys with Fe,Co,Ni,Cu and Pd two d-bands are formed with the states at the Fermi level always dominated by the Zr band. This leads to a great similarity in the electronic properties of these alloys.

Theoretical calculations of the density of states of amorphous alloys have been made and these show what we have already seen, that the crystalline and amorphous forms have very similar features. This is largely because the density of states of the d-electrons is determined by short-range interactions, that is, by the overlap of the d-wavefunctions with neighbouring ions whose type, distance and number are quite similar in the amorphous or crystalline form. The long-range order which distinguishes the crystalline state, although of great importance to the transport properties, has little effect on the density of states.

Finally, then, we ask: How are we to think of the electronic structure of a metallic glass that contains a transition element? In particular is it legitimate to picture this in k-space? Figure 8.6 shows how the Fermi surface of an alloy having an s- and a d-band might appear. The d-band might have a lot of d-character but the so-called s-band would probably be strongly hybridised with mixed s-, p- and d-character on the Fermi surface. The relative size of the two spheres would reflect the relative numbers of the two groups of electrons and their boundaries would not be very well defined.

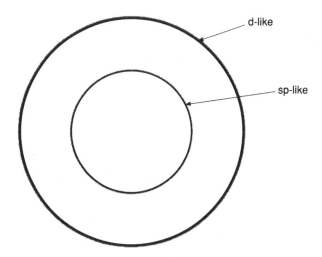

Fig. 8.6 Fermi surface of a disordered metal with s- and d-like electrons. All the states inside the d-like surface are occupied by d-like electrons and the s-like electrons are accommodated inside the s-like sphere. The surfaces will not be sharp because of the short mean free path of the electrons.

9

The Hall coefficient of metallic glasses

The Hall coefficient of simple liquid metals is, for the most part, free-electron-like. This is true, for example, of liquid Na, K, Rb, Cs, Al, Ga, In, Zn, Ge, Sn, and of liquid Cu, Ag and Au. This is also true of a wide range of non-transition metal glassy alloys of which some examples (out of many) are given in Table 9.1. Departures from the free-electron value are, however, found, for example in Ca–Al alloys and, as the table shows, in metallic glasses containing transition metal elements, which can show positive Hall coefficients in circumstances where hole conduction can scarcely be involved. To explain these positive values thus poses a problem; it emphasises still more the importance of the transition metal alloys. We must therefore look at the theory of the Hall coefficient, in particular that of alloys containing transition metals, whose behaviour forces us to recognise that the Ziman theory cannot be the whole story.

9.1 Conventional theory

Let us first consider the predictions of conventional theory for the Hall coefficient of a metal. In a crystalline metal the Hall coefficient can be difficult to calculate because it depends in a fairly complicated way on the shape of the Fermi surface and on how the electron velocities and relaxation times vary over the surface. For our purposes however we need not delve into these complexities because in amorphous metals the spherical symmetry that we assume simplifies the derivation of the Hall coefficient to the point that we can for simple alloys use the elementary expression:

$$R_H = 1/Ne \qquad (9.1)$$

Table 9.1 *Hall coefficient and other data on some metallic glasses. Liquid and crystalline Cu are included for comparison.*

Material	R_H $(10^{-11}\,\mathrm{m^3C^{-1}})$	Density $(10^3\,\mathrm{kg\,m^{-3}})$	Molar volume $(10^{-6}\,\mathrm{m^3})$	Electrons/ atom	k_F (Å^{-1})	$k_F l$	R_H [Free electron] $(10^{-11}\,\mathrm{m^3C^{-1}})$
$Cu_{50}Ti_{50}$	+10–13	7.36	10.54	1.5			−7.3
$Cu_{50}Zr_{50}$	+7.0						
$Cu_{50}Hf_{50}$	+4.0						
Liquid Cu	−8.3		7.8	1			−8.3
Cryst. Cu	−5.1		7.1	1	1.27		−7.5
$Ag_{50}Cu_{50}$	−9.1	9.75			1.27	80	−9.1
$Ca_{70}Al_{30}$	−19.6	1.85		2.3	0.98	2.3	−8.8
$Ca_{60}Al_{40}$	−15.8	1.96		2.4	1.06		−7.5
$Ca_{70}Mg_{30}$	−12.4	1.45	24.4	2	1.14		−12.4
$Mg_{75}Zn_{25}$	−6.8	2.65		2	1.4	15	−6.8
$Mg_{80.4}Cu_{19.6}$	−6.7	2.73		1.8	1.4	15	−6.7
$Sn_{86}Cu_{14}$	−4.3	7.5		3.6	1.6	16	−4.3

where e is the charge on the appropriate carrier (the electronic charge, negative for electrons, positive for holes) and N is the number of charge carriers per unit volume of the metal.

The elementary classical derivation of this result, which demonstrates the essential physics, is as follows.

We apply a magnetic field B (in the z-direction) at right angles to the direction of a current carried by a conductor. The current density is j and if the drift velocity of the electrons under the influence of the electric field alone (in say the x-direction) is δv_x, then:

$$j = Ne\delta v_x \qquad (9.2)$$

In the presence of the magnetic field B the drifting charge carriers are subject to the Lorentz force at right angles to both B and j:

$$F = Be\delta v_x \qquad (9.3)$$

This causes the charge carriers to be deflected (see Figure 9.1) so that a charge builds up on the sides of the conductor until the transverse electric field ϵ_y which results just compensates the effect of the Lorentz force. We then have:

$$e\epsilon_y = Be\delta v_x \qquad (9.4)$$

which with the use of equation (9.2) becomes:

$$\epsilon_y = Bj/Ne \qquad (9.5)$$

By comparing this with the definition of the Hall coefficient:

$$\epsilon_y = R_H Bj \qquad (9.6)$$

we see that:

$$R_H = 1/Ne \qquad (9.7)$$

In general this is positive for holes and negative for electrons although as we shall see this is not invariably so because equation (9.3) hides a subtlety which we shall explore later.

Equation (9.7) is valid if, as is implicit in the treatment, all the carriers that respond to the applied fields respond in the same way. As we have already seen, the only electrons that can respond independently to the applied fields are those at the Fermi level; the remainder are constrained by the Pauli exclusion principle to, as it were, follow their leaders. In simple metal glassy alloys there is only one set of carriers and all have similar properties because of the essential isotropy of the material. We can therefore reasonably expect the result in equation (9.7)

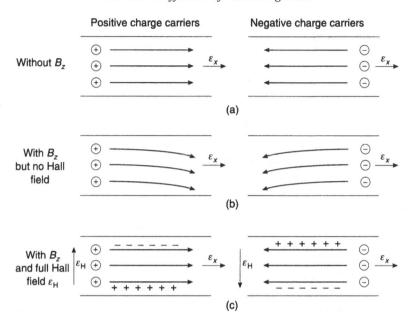

Fig. 9.1 Charge carriers of opposite signs in the Hall effect.

to be applicable and so the Hall coefficient will be independent of temperature since in a metal the number of carriers does not change.

In a transition metal alloy, however, there are, as we have seen, two types of charge carrier, the sp-like electrons and the d-like electrons. Let us therefore extend the model to include two types of charge carrier. We use the notation above with subscripts 1 and 2 to denote the properties of the two groups, remembering that the sign of the charge may also differ.

For both groups (with suitable subscripts) the current density in response to an electric field ϵ_x in the x-direction is:

$$j_x = ne\delta v_x = \sigma\epsilon_x \quad \text{so that} \quad \delta v_x = \sigma\epsilon_x/ne \qquad (9.8)$$

In the presence of the field B there is a Lorentz force:

$$F_y = Be\delta v_x = Bj_x/n \qquad (9.9)$$

This Lorentz force is different for the two groups and so, although there is ultimately no net transverse current, there are transverse currents, one associated with each group. They are in opposite directions and so cancel. The Lorentz forces cause a build-up of charge at the boundaries of the specimen, which ultimately create the transverse Hall field, ϵ_H. Let us now apply these results to the two groups of carrier for which:

$$\sigma = \sigma_1 + \sigma_2 \tag{9.10}$$

The total transverse force on electrons of group 1 is:

$$B j_{x1}/n_1 - \epsilon_H e_1$$

which is equivalent to an effective transverse electric field:

$$B j_{x1}/n_1 e_1 - \epsilon_H \tag{9.11}$$

and so the transverse current carried by the group 1 electrons is found by multiplying equation (9.11) by σ_1. The transverse current carried by the group 2 electrons is derived in a similar way. Since these must add up to zero, we get:

$$B\sigma_1 j_{x1}/n_1 e_1 + B\sigma_2 j_{x2}/n_2 e_2 - \sigma_1 \epsilon_H - \sigma_2 \epsilon_H = 0 \tag{9.12}$$

From this it follows that:

$$\epsilon_H = B[\sigma_1 j_{x1}/n_1 e_1 + \sigma_2 j_{x2}/n_2 e_2]/[\sigma_1 + \sigma_2] \tag{9.13}$$

and so:

$$R_H = \epsilon_H/B j_{xtot} = (\sigma_1^2/n_1 e_1 + \sigma_2^2/n_2 e_2)/\sigma^2 \tag{9.14}$$

since $j_{x1} = j_{xtot}\sigma_1/\sigma$ and likewise for j_{x2}.

This shows that the Hall coefficient depends on both the relative conductivity and the concentrations of the two groups of charge carriers.

A full theory of the Hall coefficient in a crystalline metal shows that, if the electron scattering rate is uniform over the whole Fermi surface, the Hall coefficient is given by:

$$R_H = (12\pi^3/e)\left\{ \left[\int (1/r)v^2 dS \right] \Big/ \left(\int v dS \right)^2 \right\} \tag{9.15}$$

The integration is over the whole Fermi surface and $1/r = (1/r_1 + 1/r_2)/2$. Here r_1 and r_2 are the principal radii of curvature of the Fermi surface at any point. If this is evaluated for a spherical Fermi surface of radius r_F we get:

$$R_H = 3\pi^2/er^3 \tag{9.16}$$

For a single type of carrier this reduces to the simple formula (9.7).

9.2 Preliminary comparison with experiment

Let us now see how far the expectations of conventional theory are met by the actual behaviour of the Hall effect in metallic glasses. The predictions of conventional theory are clear and simple when there are no d-electrons and the scattering is isotropic as expected in metallic glasses: the sign of the Hall coefficient is that of the current carriers, positive for holes and negative for electrons. It is independent of temperature. Even if there are d-electrons, equation (9.14) shows that provided that the carriers of both groups make contributions of the same sign, the sign of the Hall coefficient is that of the carriers.

As we have already seen for some of the metallic glasses made from simple metals, these predictions are found to be fulfilled but in many alloys that involve transition metals the Hall coefficient is found to be positive even though the electronic structure of the constituents makes it clear that the charge carriers, both sp-like and d-like, are electrons. Strangely enough the size of the Hall coefficient is roughly what you would expect from the sp-like electron density; only the sign is wrong. This is a very potent challenge to the theory; here, even a qualitative feature of the theory is wrong. Let us now seek an explanation of this contradiction.

Table 9.1 gives the value of the Hall coefficient of a number of metallic glasses containing transition metals. The glasses with positive values include Cu–Ti, Cu–Zr and Cu–Hf all at 50–50 composition. In these we may expect that the Cu ions will each contribute one sp-electron to the conduction band; this would be in keeping with the behaviour of Cu ions in crystalline alloys. The Ti, Zr and Hf atoms in the free state have two s-electrons and two d-electrons in their outermost shells. As we saw earlier the d-electrons contribute to a d-band associated with the Ti, Zr or Hf sites in the glass and the s-electrons form a common sp-band with the Cu ions.

It does not follow from this, however, that the numbers of s- and d-electrons will reflect those in the free atom. This depends on the relative positions of the s- and d-bands. As we saw in Chapter 8, the photo-emission spectrum of these glasses gives an indication only of the density of d-states of the corresponding alloy because the density of sp-states is masked by the d-states; the data are consistent with a d-band that is very much less than half full. There is thus no reason to suppose that hole conduction is in any way involved. Therefore the sign of the Hall coefficient should, in our interpretation, be negative.

9.3 Hybridisation of s- and d-electrons

What we have neglected and what, almost certainly, contains the key to the puzzle is the effect of hybridisation. The s-like and d-like electron states do not behave as entirely independent groups: they affect each other and dramatically alter the electron dynamics of the composite system. We have already seen how hybridisation can occur in crystalline metals and analogous effects will also occur in glasses. Indeed they will be much more pronounced because in crystals hybridisation occurs only between Bloch states of the same wave vector whereas in glasses the electron states should mix freely.

In those glassy alloys which have both s- and d-like states the E–k curve of the s-like electrons can be severely modified by interaction with the d-like states at energies where the curve passes through the d-band. There is a resonant effect analogous to that which occurs in optical dispersion when the energy (or frequency) of the incident light is close to that of a bound electron in an atom of the dispersive medium through which the light passes.

9.3.1 An optical analogy

Let us first look at this optical counterpart before considering the behaviour of the conduction electrons. When an electromagnetic wave passes through a dielectric medium, the electric field associated with the wave exerts a force on an electron in the material and, by displacing it from the centre of positive charge, creates a dipole moment, proportional to the displacement. Since, in an insulator, the electrons are bound inside an atom, the response of the bound electron depends critically on the frequency of the incident radiation. In the quasi-classical picture of the atom the bound electron is represented as an oscillator of a certain frequency ω_0, say. The aim of the theory is to calculate the dielectric constant η of the medium (from the electric dipole moment per unit volume in unit electric field) as a function of frequency. From this the refractive index $\mu = \eta^{1/2}$ and hence the velocity of propagation v can be found from $v = c/\mu$ where c is the speed of light *in vacuo*.

Figure 9.2 shows how the dielectric constant varies with frequency in this idealised model in which bound electrons of only one frequency are involved and damping effects are neglected. At low frequencies well away from ω_0, the contribution to the dielectric constant from the bound electrons is small; the motion of each in response to the electric field is limited

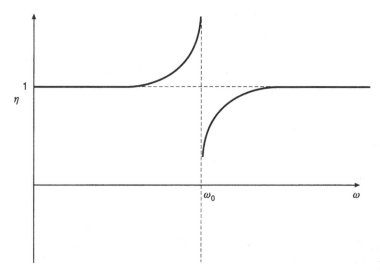

Fig. 9.2 Dielectric constant around a resonance, without damping. The change of phase from 0 to π occurs at ω_0.

and follows the direction of the field. As the frequency approaches ω_0 the tendency of the electron to oscillate at this, its natural frequency, becomes more and more pronounced until when the applied force has exactly the natural frequency ω_0, resonance occurs and the amplitude of the oscillation, and hence the dipole moment, become infinite. This is of course unphysical and, when damping is included to take account of interactions and radiation, this divergence disappears (see Figure 9.3). Above ω_0 the driving force is now trying to make the electron go faster than its natural inclination whereas below ω_0, the situation was reversed. Thus at ω_0 there is a change of phase and the sign of the contribution of the bound electrons to the dielectric constant is reversed. Thereafter as the frequency of the applied field increases and moves away from ω_0, the amplitude of the electron's motion diminishes and ultimately dies away as the inertia of the electron manifests itself in its reluctance to respond to higher and higher frequencies.

The results when damping is included are indicated in the figure which shows that the sudden reversal of the phase from 0 to π in the undamped case is now spread over the range of frequencies at which the damping is appreciable and in this range the dielectric constant and hence refractive index diminishes as the frequency rises.

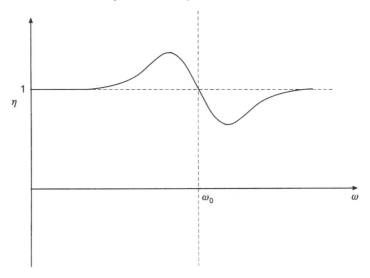

Fig. 9.3 Dielectric constant with damping. The change of phase is now spread out over the region of negative slope.

Thus, whereas well above and below the resonant frequency, the velocity of the light decreases with increased frequency, around the resonance in the region of heavy damping the reverse is true. We have 'anomalous' dispersion. This is the classical description of the phenomenon: the quantum description involves a transition of the electron between states that differ in energy by $\hbar\omega_0$ but in essence is similar to the classical argument. For comparison with the behaviour of electrons in a metallic glass, it is instructive to plot the dispersion curve as frequency ω versus wave number q as in Figure 9.4. This gives the same information as in Figure 9.3 but in a different form. In this the group velocity of the waves is given by the slope of the curve $d\omega/dq$; the phase velocity (ω/q) by the slope of the chord joining the point of interest to the origin. What is important to notice is that there is a region around resonance where the group velocity of the waves is *negative*; it occurs in the region where damping is significant.

The negative group velocity here arises from the change of phase velocity with increasing frequency. It means that the individual components of a wave packet all have positive velocities (phase velocity) but their envelope, which determines the group velocity and the total amplitude of the wave, moves in the opposite direction. There is a further point about the notion of group velocity in a spectral region where, as here, damping

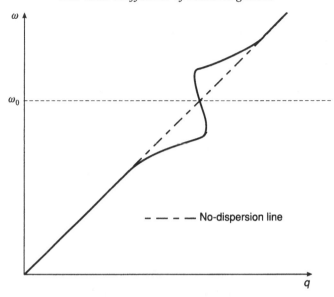

Fig. 9.4 ω–q relation corresponding to Figure 9.3.

is important. The damping is frequency dependent and so acts selectively on the components of the wave packet, which thus changes shape as it moves. The group velocity is then no longer well defined. Instead one considers the velocity with which energy can propagate or that at which a signal can be sent. It then turns out that the signal velocity cannot, for example, exceed the speed of light.

These effects are well known in optics and have been intensively studied. Let us now look at their analogue in the propagation of electrons.

9.3.2 Anomalous dispersion of electrons through s–d hybridisation

Consider first the states of s-like and d-like electrons when there is no interaction between them as shown in Figure 9.5 on an E–k diagram; this is the exact analogue of the ω–q curve just discussed with $E = \hbar\omega$. The s-like states are shown as free-electron-like, which, though an idealisation, is not too absurd since in the absence of the d-states these electrons would behave like those in a simple metal and have fairly long mean free paths. The E–k curve is thus a parabola representing the dependence $E = \hbar^2 k^2 / 2m$. The d-states are idealised to a single infinitely narrow band at energy E_d.

Figure 9.5(a) shows the situation before and after hybridisation but without damping. The consequences of the interaction between the two sets of states is quite similar to what happens in the optical dispersion we have discussed. In the electron case the wavefunction is modified so that

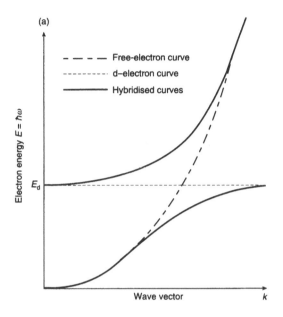

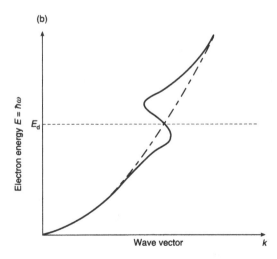

Fig. 9.5 Anomalous dispersion in s–d system. (a) Without, and (b) with damping.

the quasi-free electron whose energy is not too far from that of the d-state can be thought of as spending part of its time in that state; it is no longer a purely s-like electron but takes on some d-character and its dispersion curve is modified as shown. There are no longer separate s- and d-like electrons but hybrids of intermediate character analogous to those discussed in the section on crystalline transition metals.

In this illustration there is no damping and the d-band is unrealistically narrow. As in the optical analogue, when damping is included to take account of the scattering of the electron, the sudden phase change of π that occurs at E_d is now spread over the range of frequencies at which the damping is appreciable. This modifies the dispersion curve as before to produce a region of negative group velocity (Figure 9.5(b)). In real alloys, moreover, the d-band spans a range of energies, between E_1 and E_2, say. In addition, because of heavy scattering, the electrons in such states have quite short lifetimes so that the associated wavefunctions are heavily damped. In consequence the electron states have to be represented by their spectral functions, which can be fairly broad, and the dispersion curve is taken from the maxima of these functions. The results for a more realistic model are given in Figure 9.6, which shows that there is a region of energies in which the electron has a negative group velocity. If therefore the Fermi level of the alloy lies in this range, the electrons that determine the transport properties of the alloy will have negative group velocities and we shall now see that this reverses the sign of the Hall coefficient.

If we look at the expression for the Lorentz force on an electron of charge e and negative velocity $\mathbf{v}$ in a magnetic field $\mathbf{B}$:

$$\mathbf{F} = e\mathbf{v} \times \mathbf{B} \tag{9.17}$$

there is an alternative and formally equivalent interpretation of this equation in addition to the literal one. We can say that the force on the electron is due to a positive value of $\mathbf{v}$ and a reversed sign of e, that is, we can replace the electron by a particle with normal dynamics but a positive charge. Then the Hall coefficient is positive even though the charge carriers are negative. The formula for the Hall coefficient, equation (9.16), is still valid except that the sign of the effect is reversed. The size of the coefficient can be calculated from the radius of the Fermi sphere which now depends on both the relative positions of the s- and d-bands and the hybridisation effects.

The resistivity of the glass will also be altered by hybridisation because the size as well as the sign of the group velocity of the electrons is

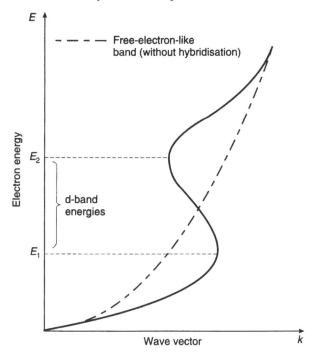

E

Electron energy

– – – – Free-electron-like
band (without hybridisation)

E_2

d-band
energies

E_1

Wave vector k

Fig. 9.6 More realistic model for s–d hybridised dispersion curve.

changed. I should, however, stress that the sign of the thermopower is not affected, a feature which will become clear when we discuss the origin and theory of thermopower in Chapter 15.

We have now some clear predictions about the sign of the Hall coefficient in metallic glasses. Where we have conduction by sp-like electrons (not holes) in a metal or alloy and the Fermi level lies within a well-defined d-band, we can expect a positive contribution to the Hall coefficient. The stress on 'well-defined d-band' is because in an alloy of two components, of which only one is a transition metal, there is, as we saw in section 8.2, a minimum concentration of this component before a d-band is established. At concentrations below this, the glass will behave like a metal with a normal dispersion relation and so have a negative Hall coefficient.

Not all the d-electrons hybridise and so we have two groups of conduction electrons. Insofar as the Fermi surface is a valid concept we must therefore think of two such spherical surfaces, one nesting inside the other. The transport properties are then derived from the contributions of these two groups, the s–d hybridised and the unhybridised d-electrons.

We must look at the consequences of this for the magnitude of the Hall coefficient. There is theoretical evidence that in some transition metals and alloys the contribution to the conductivity from the d-electrons is comparable with or greater than that of the s-electrons, presumably because the effect of their low velocities is approximately offset by their greater number. If this is so, their contribution to the Hall coefficient will be small. This follows from equation (9.14) if σ_1 and σ_2 are comparable but n_2 (the concentration of d-electrons, say) is much larger than n_1. In physical terms this is because the large number of d-electrons implies that they have a small drift velocity and so a small Lorentz force.

Model calculations confirm that the d-electron contribution may be very important to the conductivity but not to the Hall coefficient. We shall therefore for the present ignore the contribution from the d-band to the Hall coefficient, although recognising that it could make a negative contribution, partly offsetting the positive contribution of the hybridised s-electrons.

9.4 Skew scattering and the Hall coefficient

There are other mechanisms that can cause the Hall coefficient to depart from its free-electron value even when the electron properties are uniform and isotropic. One of these is skew scattering. This arises when the probability $W(\mathbf{k}, \mathbf{k}')$ of an electron being scattered by an ion from state $\mathbf{k}$ to $\mathbf{k}'$ is different from the probability $W(\mathbf{k}', \mathbf{k})$ of the inverse process. One source of such scattering is spin–orbit coupling within the scattering ion. A conduction electron that is scattered by the ion behaves while inside the ion somewhat like an atomic electron and so if the spin–orbit coupling is strong its spin couples to its orbital motion. Thus, just as the energy of the atomic electron in a state designated by $(l + \frac{1}{2})$ (where l is its orbital quantum number) can differ from one designated by $(l - \frac{1}{2})$, an electron entering the scattering field can have different energy according to its spin direction. Conversely, given its spin direction, the scattered electron takes on the orbital motion appropriate to its energy, which means that its energy and spin direction determine whether it goes round the ion in one direction or the other. Thus if you reverse the direction of the outgoing electron it will not retrace its path but go round the scatterer in the opposite sense. The two scattering events are thus not equivalent:

$$W(\mathbf{k}, \mathbf{k}') \neq W(\mathbf{k}', \mathbf{k}).$$

We write the total scattering amplitude $f_{total} = f_{ion} + f_{so}$ where f_{ion} is the scattering amplitude due to the electron–ion potential and f_{so} that due to spin–orbit coupling. The probability of scattering is then:

$$W(\mathbf{k}, \mathbf{k}') = vn_i S(q)|f_{total}|^2 \qquad (9.18)$$

where v is the electron speed, $S(q)$ the structure factor and n_i the number of scatterers per unit volume. As before $q = 2k_F \sin(\theta/2)$ where θ is the scattering angle. From the definition of f_{total} already given, we see that:

$$|f_{total}|^2 = (f_{ion})^2 + 2f_{ion}f_{so} + (f_{so})^2 \qquad (9.19)$$

The first term is large and gives rise to the usual ionic scattering, the second gives rise to the skew scattering and the third produces a change of spin direction but is a second-order effect. The second or cross term concerns us. It involves the factor:

$$\sigma_z \cdot \mathbf{k} \times \mathbf{k}' \qquad (9.20)$$

where σ_z is a unit vector in the field direction, $+$ for spin up and $-$ for spin down. The vector product shows that interchanging $\mathbf{k}$ and $\mathbf{k}'$ changes the sign of this term in the scattering probability and is thus the source of the skew scattering. Other factors in the cross term, which are not included since they influence only the magnitude and not the mechanism of the effect, are the Pauli susceptibility of the electrons and the spin–orbit parameters. These parameters obviously depend on the angular momentum of the electron; for example, they are zero for s-electrons for which $l = 0$. They are different in detail from the atomic parameters because of the different boundary condition on the electron wavefunction when the ion is in a solid.

The factor given by equation (9.20) enables us to see how the Hall effect comes to be altered. We can interchange the dot and the cross in equation (9.20) so that the first two factors involve the vector product of a unit vector in the spin direction and a vector in the direction of the electron motion. Since, in the magnetic field which is present in the Hall effect, there are more spins pointing along the field direction than against it, we consider an electron with spin in this majority direction, that of $\mathbf{B}$. We also consider a k-vector in the electron current direction $\mathbf{j}$ because a majority of electrons must possess components in this direction to sustain the current. Thus equation (9.20) is proportional to:

$$\mathbf{B} \times \mathbf{j} \cdot \mathbf{k}' \qquad (9.20a)$$

Because in the Hall effect we measure the potential transverse to the field and current direction, we are concerned only with $\mathbf{k}'$ vectors normal to $\mathbf{B}$ and $\mathbf{j}$. Thus the expression (9.20a) varies as $\mathbf{B} \times \mathbf{j}$ which has the same angular dependence as the Lorentz force. In this way it contributes to the Hall coefficient.

The final magnitude of the spin–orbit effect depends on combining equation (9.20) with the structure factor and the scattering amplitude of the electron–ion potential, both of which usually depend on scattering angle. Since, however, they do not depend on $\mathbf{B}$, the angular dependence of the result is not altered and the skew scattering makes a true contribution to the Hall coefficient.

The effects of spin–orbit interaction are thought to explain the reduction in the Hall coefficient of liquid mercury when it forms alloys. Pure liquid mercury, in spite of strong spin–orbit coupling, has a Hall coefficient quite close to the free-electron value and this is thought to arise because the angular dependence of the other factors mentioned above fortuitously annul the spin–orbit contribution. The alloying additives which all reduce the Hall coefficient have one thing in common, namely, a small spin–orbit parameter relative to mercury. The general idea is that the masked effect of the strong spin–orbit coupling in mercury itself is revealed when the fortuitous cancellation is removed on alloying.

Spin–orbit effects tend to be confined to the metals of high atomic number such as gold, platinum and mercury, because the spin–orbit coupling parameter increases very markedly with atomic number. It seems clear that skew scattering cannot be the explanation of the positive Hall coefficient in so many transition metal glasses made from elements that are of comparatively low atomic number.

There is one further point about spin–orbit coupling that will be of use to us later: because of this coupling the spin is no longer a good quantum number when the electron is inside the ion. Spin–orbit coupling can thus disorient the spin in the scattering process and randomise its direction. This has important consequences for the phase of the electron wavefunction.

There are other effects such as side-jump scattering that have been invoked to explain the anomalous Hall coefficient in ferromagnetic metals and considered as possible explanations of the sign change of the Hall coefficient that we have been discussing. Their origin and applicability are controversial and will not be discussed here but references to publications where they are discussed are given in the notes[1].

We now turn to a comparison of experimental results with the theoretical ideas that we have been discussing.

9.5 Experimental results

When the Hall coefficients of metallic glasses were first measured they were thought to be independent of temperature but later, more careful measurements showed that there was a slight temperature dependence. This is a subtle effect whose origin we shall explore later when we have studied the nature of scattering processes in these materials more thoroughly. In the meantime we ignore this slight dependence on temperature and generally quote room temperature values.

Figure 9.7 provides a way of comparing the Hall coefficient for a range of noble metal alloys and of pure metals in the amorphous state, some liquid, some glassy. R_H is plotted against alloy concentration after normalisation to the free-electron values, calculated by assuming that each atom contributes one electron to the conduction band. At low concentrations of the noble metal, where the open d-band would not be expected to exist, the Hall coefficient is negative but where the d-band is securely formed the sign is positive right up to the pure liquid. These findings accord with the anomalous dispersion theory.

In Table 9.2 we see values for a number of metallic glasses in the series of Cu with Ti, Zr and Hf. Liquid and crystalline Cu are included for comparison. The table demonstrates again the change in sign of the Hall coefficient when the proportion of Cu becomes high. (It was found impossible to make a glass of $Cu_{90}Ti_{10}$.) In alloys rich in Ti, Zr or Hf values of R_H not quoted in the table do not vary much from those given for the 50–50 alloys. Presumably this is because, once the non-copper element dominates, the Fermi level is determined by the density of states in the open d-band and so does not change much. However, as the proportion of Cu increases beyond about 65 %, R_H declines before changing sign. This, like the similar effect in Figure 9.7, may reflect the weakening of the hybridisation before the open d-band ceases to exist.

There is clearly a systematic change in the Hall coefficient of the concentrated alloys as we go from Ti to Zr to Hf but its origin is not clear. It could be due to a change in the band structure, for example, the position and width of the open d-band. In the crystalline transition metals, the d-band tends to widen and sink in energy relative to the s-band as you go from the 3d to 4d to 5d. If this occurred here and the s-band remained unchanged, it would suggest a systematic reduction in k_F for the hybridised electrons and so, contrary to what is observed, an increasing Hall coefficient (see equation (9.14)). The systematic change could be due to an increasing contribution from the d-electrons but, as we saw earlier, the

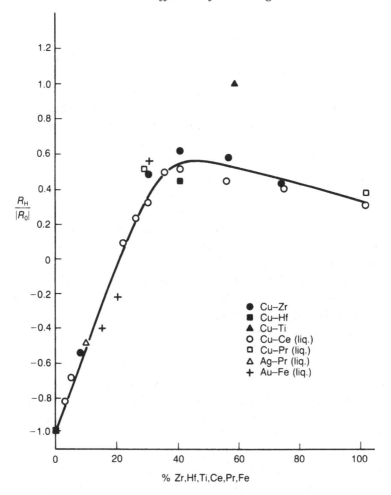

Fig. 9.7 Normalised Hall coefficient versus electron concentration. R_H is the measured Hall coefficient and R_0 the free-electron value. (After Weir *et al.* 1983.)

theoretical models suggest that this contribution should be small whereas the changes to be explained are substantial.

The change is unlikely to be related to spin–orbit effects since these are small in ions of the atomic numbers involved here. At present we do not have enough detailed knowledge to explain the trend. It would be instructive to see how R_H changes as we move the Fermi level through the region of anomalous dispersion. To do this one would like measurements on alloys of Cu with, say, the 4d series Y,Zr,Nb,Mo,(Tc),Ru and Rh. Apart

Table 9.2 *Hall coefficient of some copper-based metallic glasses*

Alloy	Hall coefficient $(10^{-11}\,\mathrm{m}^3\,\mathrm{C}^{-1})$
$Cu_{50}Ti_{50}$	$+10\text{–}13$
$Cu_{50}Zr_{50}$	$+7.0$
$Cu_{50}Hf_{50}$	$+4.0$
$Cu_{68}Ti_{32}$	$+6.5$
$Cu_{70}Zr_{30}$	$+4.6$
$Cu_{70}Hf_{30}$	$+2.2$
$Cu_{90}Ti_{10}$	
$Cu_{90}Zr_{10}$	-5.1
$Cu_{90}Hf_{10}$	-5.2
Liquid Cu	-8.3
Cryst. Cu	-5.1

from Y and Zr, it has so far proved impossible to make and measure such samples.

Other alloy series that have been systematically measured and which show a change with concentration of the sign of the Hall coefficient are NiZr and CoZr. Liquid Ni and liquid Co both have negative values as do the glassy alloys rich in these elements. Where the Zr is dominant the values are positive. In these alloys both elements contribute d-electrons to the conduction process.

The explanation of the positive Hall coefficient in some metallic glasses is a key problem. The s–d hybridisation theory seems to be well founded[2] and, although not perhaps universally accepted, appears to be the most plausible.

10

Magnetoresistance

10.1 Qualitative picture

First of all we concentrate on the transverse magnetoresistance in which the magnetic field is applied normal to the current direction. The calculation of the magnetoresistance of a crystalline material is very difficult unless there are simplifying features. In the metallic glasses fortunately there are indeed such features. If we make the same assumptions as in our first derivation of the Hall coefficient we find zero magnetoresistance. The effect of the magnetic field is so perfectly compensated by the transverse electric field (the Hall field) that the resultant current is completely unperturbed and so there is no change in resistance i.e. no magnetoresistance.

In the alloys of non-transition metals there is only one type of charge carrier and no obvious source of anisotropy so the magnetoresistance due to conventional mechanisms must be vanishingly small.

If there is to be a non-zero magnetoresistance some additional feature has to come into the story. One example of such a feature is the presence of the two different types of charge carrier that we postulated for transition metal alloys.

10.2 Two-band model

If we assume that there are two kinds of carrier, we can perhaps understand the physics of this type of magnetoresistance in macroscopic terms as we did for the Hall effect. The first effect of the magnetic field is, as we saw, to produce a transverse component to the electric current; in the normal geometry of resistance measurement this current is suppressed by the boundaries of the specimen: the transverse Hall field builds up until

100

there is no net transverse current. If there are two types of carrier each type has a different drift velocity in the electric field and thus experiences a different Lorentz force. Suppose that group 1 have a low drift velocity and group 2 have a high one. Clearly a transverse electric field cannot exactly balance both Lorentz forces; to suppress the transverse current the field builds up to a value which overcompensates the weaker Lorentz force and undercompensates the stronger one. When we take account of the combined effect of the Lorentz force and that due to the Hall field, both of which act in a direction transverse to the current direction, it is clear that the group 1 electrons will produce a transverse current in one sense and group 2 an equal current in the other.

The magnetic field will now act on the resulting transverse current components to produce additional currents at right angles to their transverse direction. The current which suffered the greater Lorentz force in the first instance will do so again and this time will be turned through another right angle so as to run counter to the original current direction; it will thus reduce the total current and so tend to increase the resistance to flow. The transverse component in the opposite direction will suffer a Lorentz force that moves it into the direction of the main current thus tending to reduce the resistance. It turns out that the net effect i.e. the difference between the additional currents flowing with and against the main stream is always in the opposite sense to the primary current and so always increases the resistance (for the two-group model just described this can be demonstrated straightforwardly). Thus the magnetoresistance defined as $\Delta\rho(B)/\rho(B=0)$ due to this mechanism is always positive.

The above argument envisages two stages to the evolution of the magnetoresistance: first the establishing of the transverse Hall field, which is proportional to B, and then the operation of B again on the resulting transverse components of the current to produce the final change in the applied current which causes the resistance change $\Delta\rho$. This second stage is also proportional to B at low fields and so the magnetoresistance under these conditions is proportional to B^2. The magnetoresistance calculated for this two-group model (in the same notation as in Chapter 9) is:

$$\Delta\rho/\rho(B=0) = \sigma_1\sigma_2[(\sigma_1/n_1e_1) - (\sigma_2/n_2e_2)]^2 B^2/\sigma^2 \qquad (10.1)$$

Here the electronic charge e is to be thought of as having the appropriate sign.

The effect we have so far discussed is the transverse magnetoresistance but there is also a longitudinal effect observed when the magnetic field is directed along the direction of the current. This vanishes in the two-group model we have just discussed. Clearly this is a more subtle effect since there is no Lorentz force on an electron travelling parallel to the magnetic field. We must however remember that in general the electron velocities are not all equal and are not parallel to the current direction although of course they have components in that direction. These components are unchanged by the longitudinal magnetic field but the components normal to this direction are altered by it and would cause the electrons to spiral around the field direction if they were not scattered. Where there are sources of anisotropy this longitudinal magnetoresistance is also positive and proportional to B^2 at low fields.

10.3 Kohler's rule

The magnetoresistance that I have just described is often referred to as the Kohler magnetoresistance after the theorist who first identified its main features. Kohler also devised a criterion known as Kohler's Rule to determine whether this type of magnetoresistance is likely to be large or small.

A dimensional argument allows us to work out a useful criterion analogous to Kohler's Rule. If an electron of mass m and charge e moves at right angles to a magnetic field B with angular velocity ω, it describes a circle of radius r given by:

$$mr\omega^2 = evB \qquad (10.2)$$

where v is the linear speed given by $v = r\omega$. So the so-called 'cyclotron frequency' of the electron is:

$$\omega_c = eB/m \qquad (10.3)$$

The reciprocal of this frequency is of the order of the time the electron takes to execute one orbit in the field B. We can compare this with the mean time τ between collisions to see how much effect the magnetic field has before the electron is scattered out of its orbit and its direction randomised. The dimensionless quantity $\omega_c\tau$ is a measure of this; it is the angle through which the radius vector of the electron in its cyclotron orbit turns before the electron is scattered. If $\omega_c\tau \ll 1$, it means that the electron is scattered almost as soon as it begins its orbit; it is deflected through a very small angle and so the effect of the field B is very small. Since we know that the low-field magnetoresistance is proportional to B^2 and that ω_c is proportional to B, we can get a rough measure of the likely size of the low-field magnetoresistance $\Delta\rho/\rho$ by taking it as of order $(\omega_c\tau)^2$.

Since in the simplest model, we can write $\sigma = ne^2\tau/m$, $\omega_c\tau$ can then be expressed as $\sigma B/en$.

10.4 Rough estimates

Let us make a crude estimate of the magnetoresistance in a simple two-band model with s- and d-electrons. If we assume that half the current is carried by the d-electrons (say) and that they constitute 80 % of the number of carriers, the magnetoresistance, according to equation (10.1) works out at about $(\omega_c\tau)^2$. The contribution of the d-electrons is small because their number is relatively large and their velocity small; this tends to make their drift velocity small (in a given electric field) and so make the Lorentz force and hence the d-contribution to the Hall field and magnetoresistance small. In this calculation I have also assumed that the s-electrons have an effective charge opposite in sign to that of the d-electrons (because of the anomalous dispersion) so that the two terms in the square brackets of equation (10.1) add up. The calculation is then not sensitive to the precise numbers, which it could be if the two terms in the square brackets were of opposite sign. In that case of course the magnetoresistance would be smaller.

To get an idea of the magnitudes involved we note that according to equation (10.3) the cyclotron frequency of a free electron in a field of one tesla (1 T) is $1.6 \times 10^{-19} \div 9.1 \times 10^{-31} = 1.76 \times 10^{11}\,\text{s}^{-1}$. So in any material in which τ is of order 10^{-13} s or less we can expect the magnetoresistance in such a field to satisfy the condition $\omega_c\tau \ll 1$; in typical alloys with which we are concerned where τ is about 10^{-16} s the Kohler magnetoresistance in magnetic fields of 10 to 20 T is going to be completely negligible (of order 10^{-7}).

Measurements show, however, that at low temperatures the magnetoresistance in some of these glasses is substantial, much bigger than predicted by the Kohler criterion and quite easily measurable. Moreover, although it varies as B^2 at very low fields, it changes in some samples to a $B^{1/2}$ dependence at higher fields, which suggests that something unusual is involved. In other samples and circumstances, the behaviour is a good deal more complex. Here then is another mystery to be explained.

The brief survey in this and the previous chapter has shown that although the Boltzmann-type theories that we have already looked at can give us some insight into the interpretation of the electronic properties of metallic glasses, there are some outstanding features that demand the revision or extension of our theoretical picture. We now look at what these extensions are and how far they explain the details of the resistance and magnetoresistance of metallic glasses.

11

Electrical conductivity of metallic glasses: weak localisation

Having looked at some of the ideas in terms of which the electrical conductivity of metals has conventionally been interpreted, we now look at the conductivity of metallic glasses to see how far we can understand it in terms of what we have learned. The broad features of the conductivity of glasses made from simple metals have been interpreted in terms of the Ziman model (as established for simple metal liquids). Those that contain a substantial proportion of at least one transition metal have properties that cannot, for the most part, be so interpreted and indeed it was soon recognised that even simple metal alloys require an extension of the theory. Because all these materials we are considering are highly disordered, we can be sure that their electrical resistivity will be large at all temperatures and will not vary a great deal with temperature; its precise magnitude will of course depend on the specific constituents of the alloy.

There is one generalisation that can be made at the outset. Experimental data show that, as we would expect, the residual resistivity, ρ_0, of a glass is comparable to that of the corresponding liquid and indeed its resistance looks like the natural continuation of that of the liquid to low temperatures. This is illustrated in Figure 11.1 for $Ni_{60}Nb_{40}$ and $Pd_{81}Si_{19}$, which also shows that the crystalline form at low temperatures with its much higher degree of order has a much lower resistivity. All this is reassuring.

Let us first consider the alloy $Cu_{60}Ti_{40}$ whose resistivity is about $190\,\mu\Omega\,cm$. What does this tell us about the mean free path of the conduction electrons in this alloy? To get a rough idea we assume that the conduction electrons are free-electron-like and use the relation $\sigma = ne^2\tau/m$. To estimate n we suppose that each Cu atom and each Ti

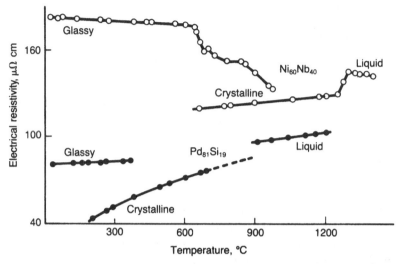

Fig. 11.1 The resistivity of alloys, $Ni_{60}Nb_{40}$ and $Pd_{81}Si_{19}$ in liquid, glassy and crystalline form as a function of temperature. Notice that the resistivity of the glassy phase appears to form a reasonable extrapolation of that of the liquid. (After Güntherodt *et al.* 1978.)

atom contribute one electron to the s-band and hence to the conduction process; for our present estimate we ignore the contribution of the d-electrons. From n we can then calculate k_F to be $1.3 \times 10^8 \, \text{cm}^{-1}$ and the Fermi velocity v_F to be about $10^8 \, \text{cm s}^{-1}$. From the relation $l = v_F \tau$ we find for the mean free path a value of about 4×10^{-8} cm, which is little more than the interionic spacing. With such a short mean free path we can no longer maintain our picture of conduction electrons travelling distances long compared to both the interatomic spacing and the Fermi electron wavelength, with just occasional scattering processes to interrupt their progress. This is only part of the problem.

11.1 The negative temperature coefficient of resistivity

The temperature coefficient of resistivity $\alpha = (1/\rho) \, d\rho/dT$ of most glasses is small as we would expect but the resistivity of many glasses falls slightly with rising temperature, which we would not expect. This is a comparatively rare phenomenon in crystalline metals and is not explained by Boltzmann theory unless there are special circumstances. Moreover there is an interesting correlation: those alloys with ρ_0 less than $150 \, \mu\Omega$ cm tend to have positive temperature coefficients (around room temperature) whereas those with ρ_0 greater than $150 \, \mu\Omega$ cm tend to have

Table 11.1 *The resistivity and its temperature coefficient in some metallic glasses and liquid iron*

Material	Resistivity, ρ ($\mu\Omega$ cm)	$(1/\rho)\mathrm{d}\rho/\mathrm{d}T$ (10^{-4} K^{-1})
$Cu_{50}Zr_{50}$	178	−1.0
$Cu_{50}Ti_{50}$	204	−1.0
$Cu_{60}Hf_{40}$	190	−1.2
$Ni_{20}Zr_{80}$	160	−0.9
$Ni_{50}Zr_{50}$	184	−1.4
$Ni_{70}Zr_{30}$	164	−0.3
$Ni_{50}Nb_{50}$	195	−0.08
$Co_{30}Zr_{70}$	173	−1.3
$Fe_{33}Zr_{67}$	168	−1.0
Liquid Fe	135	+2.0
$Pd_{80}Si_{20}$	102	+0.7
$Pd_{80}Ge_{20}$	101	+1.87
$Ni_{80}P_{20}$	130	+0.6

negative ones. This is known as the Mooij correlation after its discoverer (Mooij 1973) and is illustrated in Table 11.1 and Figure 11.2.

Metallic glasses with $\rho_0 > 150\,\mu\Omega$ cm often show a resistance minimum, i.e. as we start from low temperatures the resistance first falls and then rises. The fall in resistivity in those metallic glasses that have negative temperature coefficients is small, typically a few per cent between low temperatures and room temperature. Nonetheless this is important because in all our discussions of phonon scattering the effect of increasing the temperature has been to increase the disorder in the system and so to increase the resistivity.

A well-established cause of a negative temperature coefficient is the presence of magnetic scatterers. One example is the Kondo effect, which occurs in very dilute alloys containing impurities that carry a magnetic moment in solution. Another example is found in ferromagnetic plutonium in the neighbourhood of the Curie point.

A negative temperature coefficient of resistivity can also arise in metals or alloys which have a very sharp drop in the density of states near to the Fermi level, an effect seen for example in some alloys of palladium. It is, however, unlikely to be present in glasses where the disorder usually smears out any such sharp features.

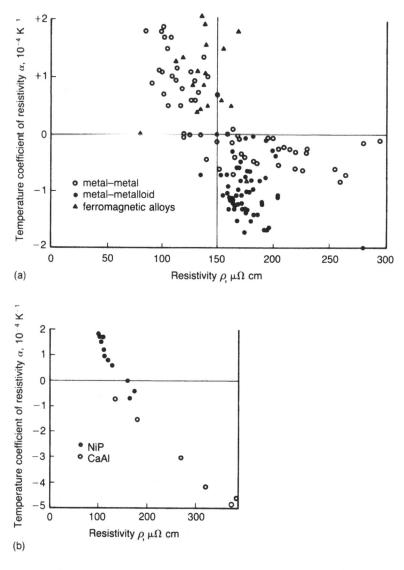

Fig. 11.2(a) The temperature coefficient of resistivity α versus resistivity ρ for a range of metals and alloys to illustrate the Mooij correlation. (b) The same plot for two particular alloys. (After Howson and Gallagher 1988.)

We have also seen how changes in the structure factor in some monovalent liquid metals can cause the resistivity at high temperatures to fall with rising temperature.

All these are however rather special conditions that do not apply in most of the metallic glasses that concern us. Thus, whereas the generally high value of the resistivity is understandable, the Mooij correlation and the details of the temperature dependence are not. These matters ultimately demand an extension of conventional theory.

The position can be summarised in the following way. Hitherto we have considered the Fermi electrons as travelling large distances (compared to their wavelength and the interatomic spacing) without being scattered. The scattering events could thus be treated quite independently and the structure factor, which describes the consequences of single scattering processes, is sufficient to account for this aspect of the scattering. Now, however, we are confronted with scattering processes so frequent that we are forced to consider the possibility of interference between incident and scattered waves. Moreover the electron progresses from ion to ion by almost a random walk so that we are led to think of its motion as diffusive rather than ballistic. In the detailed treatment that follows, the classical diffusion equation allows us to treat interference between the scattered waves (neglected in the Boltzmann approach) at least in an approximate way and to achieve a remarkable insight into the behaviour of these electrons.

11.2 Interference effects

The conduction electrons propagate through the metallic crystal or glass as waves and, where interference is possible, its consequences depend on the relative phase of the interfering waves. If a wave of wavelength λ moves from point A to point B its phase changes and if A and B are in the line of propagation and are one wavelength apart the difference in phase of the wave at A and at B is 2π. Correspondingly if the wave goes a distance dq in the direction of propagation its phase changes by $2\pi dq/\lambda$ or, in terms of the wave number $k = 2\pi/\lambda$, by kdq; from the de Broglie relation $p = \hbar k$, where p is the momentum of the electron, this phase change can be written as $pdq/\hbar$. If the direction dq is not that of p or k, we can use vectors and write $\mathbf{k} \cdot \mathbf{dq}$ or $\mathbf{p} \cdot \mathbf{dq}/\hbar$. For a specific path the total phase change is obtained by integrating along it. Notice that the value of $\mathbf{p} \cdot \mathbf{dq}$ integrated over the path is just the classical action A associated with the path. So the total phase change along the path is $A/\hbar$.

Since the phase of the wavefunction is so important in what follows, let us digress slightly to see how its connection with the Principle of Least Action links classical and quantum mechanics.

When a quantum particle travels from A to B, the path it follows is determined by the relative probability of all possible paths between A and B. But these probabilities are expressed as amplitudes (dependent on phase) and not intensities. The ultimate behaviour is then determined by finding the resultant probability amplitude and taking the square modulus to find the final probability intensity. Let us suppose that the path and object are macroscopic so that the number of de Broglie wavelengths contained in the path is enormous. Any arbitrary path therefore has close to it other paths whose phase is quite different to its own because a tiny change in path length produces a big change of phase. So in general these paths tend to cancel each other. Only in the neighbourhood of a path for which the phase and hence the action is an extremum do the neighbouring paths, differing only infinitesimally from the extremum, have phases that to first order are the same as that of the original path. The associated probability waves are therefore in phase and reinforce each other. Thus the only path that survives with appreciable quantum probability is that associated with an extreme value of the action; this is just the classical path determined by the *Principle of Least Action*.

Because we have to deal with electrons in a magnetic field, let us determine how the phase of the electron wavefunction is altered when exposed to a magnetic field *B*. In writing down the phase change in the absence of a field we used the so-called 'canonical coordinates' *p* and *q* of the Hamiltonian formulation of classical mechanics. This then must be our guide in extending the expression to include a magnetic field. To do this in classical mechanics we replace the momentum $\mathbf{p}$ by $(\mathbf{p} - e\mathbf{A})$ where *e* is the electronic change and $\mathbf{A}$ (not to be confused with the action *A*) is the vector potential corresponding to $\mathbf{B}$ i.e. $\mathbf{B} \equiv \text{curl} \, \mathbf{A}$. Thus in the corresponding quantal system the phase change associated with a displacement $d\mathbf{q}$ is $(\mathbf{p} - e\mathbf{A})d\mathbf{q}/\hbar$. To find the change of phase when the electron executes an arbitrary closed path O, we must integrate the above expression for the phase around the path O. We have two terms: the first is just the path integral of $\mathbf{p}/\hbar$ which we have already discussed and which we take to be independent of the magnetic field. The second is proportional to $\mathbf{A} \cdot d\mathbf{q}$ integrated around O. We can, however, rewrite this since:

$$\oint \mathbf{A} \cdot d\mathbf{q} = \int \text{curl} \, \mathbf{A} \cdot d\mathbf{S} \qquad (11.1)$$

where $d\mathbf{S}$ is an element of area and the integral on the right-hand side is over any surface bounded by O. Since $\mathbf{B} \equiv \text{curl} \, \mathbf{A}$ we see that the right-hand side of equation (11.1) is $\int \mathbf{B} \cdot d\mathbf{S}$ which is the total flux of $\mathbf{B}$ through the closed path O. If we call this flux ϕ then the phase change induced by the field $\mathbf{B}$ when the electron traverses the path O is $e\phi/\hbar$. Thus we see that the change of phase induced by the field is just equal to the change in flux through the circuit expressed in units of $e/\hbar$.

11.3 The Aharonov–Bohm effect

The result just stated has long been known but it acquired additional interest and importance when Aharonov and Bohm (1959) predicted that if a charged particle encircled a region of magnetic field it would suffer a change of phase in proportion to the magnetic flux enclosed by the path *even though the particle never itself entered the magnetic field*. This was at once verified with electrons in a very elegant experiment by Chambers, using an electron microscope as the source of the electrons and an iron whisker to confine the magnetic flux. In this way he could divide the electron beam to pass outside the whisker and then recombine it to form an interference pattern which he could use to measure the phase shift.

These results were achieved with electrons in free space. We now turn to analogous work with conduction electrons in metals. Altshuler, Aronov and Spivak (1981) predicted that the Aharonov–Bohm effect just described should be observable with conduction electrons in a metal at low temperatures. The prediction was verified by Sharvin and Sharvin (1981) in the following experiment.

They took a thin-walled cylinder of lithium metal at low temperatures inside which they could generate a magnetic field parallel to the axis of the cylinder but totally confined inside the cylinder with no stray field in the metal itself (see Figure 11.3). They then measured the resistance of the cylinder between its ends as a function of the interior magnetic field and found that the resistance showed oscillatory variations of period $\hbar/2e$ in the flux as illustrated in Figure 11.4.

Why does this happen? The main effects arise because of interference in a special class of electron trajectories, namely those that contain closed loops. This interference comes about because an electron can execute such a closed path in two different senses as illustrated in Figure 11.5. In fact there is an additional condition that must be satisfied: the scattering processes that produce the closed path must all be elastic so that phase coherence is maintained throughout; scattering at low temperatures by the static disorder is in general elastic with no phonons involved.

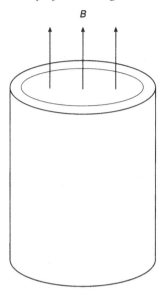

Fig. 11.3 Thin-walled cylinder with magnetic flux confined within the inner wall. The conduction electrons never enter the magnetic field itself.

Any inelastic scattering event would introduce an arbitrary phase change and destroy the interference effects.

Let us now calculate the probability W of an electron leaving and returning to the point X on a closed path. If ψ_1 is the probability amplitude of the anticlockwise path and ψ_2 the probability amplitude of the clockwise path, then the total probability of the path is:

$$W = \left| \sum \psi_i \right|^2 \qquad (11.2)$$

where the sum is over all possible paths. Here this is:

$$W = |\psi_1|^2 + |\psi_2|^2 + \psi_1^* \psi_2 + \psi_1 \psi_2^* \qquad (11.3)$$

The first two terms are just classical probabilities and the second two terms are interference terms that would not appear in a classical argument. Even in a quantal argument they would generally average to zero when summing over all possible paths since there would usually be arbitrary phase differences between the contributions to ψ_1 and ψ_2. In the particular case considered here however the two paths are of identical length and involve identical scattering events although encountered in the reverse sequence. Moreover phase coherence is maintained throughout.

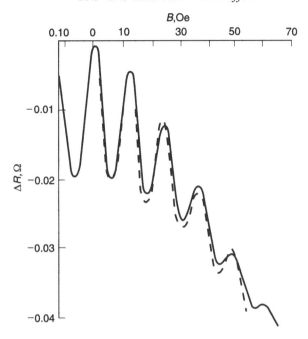

Fig. 11.4 Change in resistance of hollow cylinder versus magnetic field showing its periodic variation with the flux. (After Altshuler *et al.* 1982.)

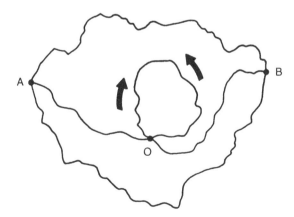

Fig. 11.5 Possible electron paths including a closed path in which an electron returns to its starting point. On the closed path there are two paths of equal length corresponding to the two senses in which the path can be traversed; these are indicated by the arrows.

Thus the probability amplitudes ψ_1 and ψ_2 associated with the two paths are completely in phase and all four terms on the right-hand side of equation (11.3) have the same value: for these special paths W has twice its classical value. This means that processes in which an electron returns to its starting point after a succession of elastic scattering events have a probability enhanced above that expected in a classical calculation and the electrical resistivity of the cylinder is enhanced above its 'classical' value by this quantum interference. Moreover, in some of these special processes the electron travels right round the cylinder before returning to its starting point, as indicated in Figure 11.6. Let us concentrate our attention on these.

When a magnetic field is applied there is a change in the magnetic flux through these paths. Call this $\Delta\phi$. This, as we have seen, alters the phase of the associated wavefunction by $e\Delta\phi/\hbar$. The effect of this is of one sign for ψ_1 and of the opposite sign for ψ_2 since it describes an electron executing the path in the opposite sense. There is thus a relative phase shift of $2e\Delta\phi/\hbar$ between the two and as the field, and hence the flux ϕ, is increased from zero the two wavefunctions change from being exactly in phase and become more and more out of phase. The enhanced resistance due to quantum interference thus gets smaller and when the waves are in antiphase the enhancement of the resistance is completely destroyed. Thereafter as the field increases further there will be a periodic variation in resistance as the flux through the cylinder changes. The period of this change is $\hbar/2e$ exactly as observed in the experiment.

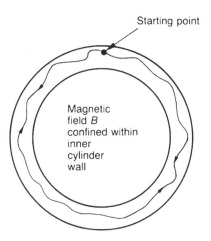

Starting point

Magnetic
field B
confined within
inner
cylinder
wall

Fig. 11.6 Electron closed path that goes right around the cylinder.

11.4 Weak localisation

This remarkable result demonstrates beyond doubt the existence of quantum interference effects in the behaviour of conduction electrons. We shall now see how these effects influence the behaviour of metallic glasses (or indeed any highly resistive metal) when the geometry of the sample is of conventional form.

The electron paths that interest us here are those that contain closed loops. As we saw they have a probability enhanced by quantum interference above that of non-intersecting paths. This means that the chance of an electron returning to its starting point is increased and because of this the phenomenon is often referred to as 'weak localisation'. It also means that the material will show a higher resistivity than would be expected from theories that neglect this enhanced scattering.

We can now understand why the resistance of a metallic glass falls as the temperature rises. Let us start with the specimen at absolute zero where inelastic scattering is at a minimum. The resistance is thus enhanced by the tendency to weak localisation that we have just discussed. Now we raise the temperature gradually; inelastic scattering becomes more probable as, for example, more and more phonons are excited and this inelastic scattering destroys the coherence of the wavefunction of some of the electrons that were participating in the enhanced scattering. This in turn destroys the quantum interference in these loops, reduces the probability of back-scattering and so reduces the resistance. Thus we expect to see the resistivity fall with rising temperature. We can now understand the two features of the resistivity of metallic glasses that I have already mentioned: first, why the temperature coefficient is negative and second why this negative temperature coefficient is correlated with high resistivity.

We are able to take the argument a stage further if we assume that the electron is executing a random walk from scattering site to scattering site and that the probability of its return to its starting point can be calculated on the basis of classical diffusion theory. In three dimensions the probability per unit volume that a particle leaving the origin at time $t = 0$ should be at a distance r at time t is:

$$p(r, t) = [\exp(-r^2/4\pi Dt)/(4\pi Dt)^{3/2}] \tag{11.4}$$

(This is valid in three dimensions; in one or two dimensions the result is different and leads to different behaviour for the conductivity.) D is the diffusion coefficient of the conduction electrons and can be deduced from the electrical conductivity using the Einstein relation (equation (3.41)).

To find the probability of the electron returning to its starting point we put $r = 0$ in the above expression and estimate the volume of space occupied by the electron in an element of time dt. We may think of the electron as having the dimensions of its wavelength λ and travelling with the Fermi velocity v_F. In a time dt it therefore sweeps out a volume $\lambda^2 v_F dt$. The probability that its path will return to the origin and find itself in this volume is thus approximately:

$$p(t)dt \simeq \lambda^2 v_F dt/(4\pi Dt)^{3/2} \qquad (11.5)$$

Now we need the value of this probability integrated over the period during which the electron wavefunction remains coherent. If τ_{in} is the inelastic scattering time after which, on average, the coherence of the electronic wavefunction is destroyed by an inelastic scattering event, we must integrate equation (11.5) from the minimum time of a scattering process τ_0 (the mean free time for elastic scattering) to τ_{in}. This then yields the probability of enhanced back-scattering as:

$$\int_{\tau_0}^{\tau_{in}} p \, dt = p(\tau_{in}) \simeq \lambda^2 v_F[(\tau_0)^{-1/2} - (\tau_{in})^{-1/2}]/(4\pi D)^{3/2} \qquad (11.6)$$

This reduces the value of the diffusion coefficient D by the same fraction and we know from the Einstein relation (equation (3.41)) that:

$$\sigma = e^2 D N(E_F) \qquad (11.7)$$

Thus if D is reduced (without change in the density of states) σ is reduced in proportion. Since the change is small we can write for the reduction in conductivity $\Delta\sigma$ and the enhancement of the resistivity $\Delta\rho$:

$$\Delta D/D = \Delta\sigma/\sigma_0 = -\Delta\rho/\rho_0 = -p(\tau_{in}) \qquad (11.8)$$

If now in equation (11.6) we write $D = v_F l/3$ and $\lambda = 2\pi/k_F$, we find:

$$\Delta\rho/\rho_0 = -\Delta\sigma/\sigma_0 = \alpha[1 - (\tau_0/\tau_{in})^{1/2}]/(k_F l)^2 \qquad (11.9)$$

where α is a numerical constant, shown by a full calculation to be 3.

We see from this that when the inelastic scattering time equals that for elastic scattering $\Delta\rho$ or $\Delta\sigma$ vanishes, as we would expect. Moreover the temperature dependence is determined by that of the inelastic scattering time, as seen in the second term of the square bracket. The size of the effect is determined by the quantity $(k_F l)^2$; if this is large, so that the mean free path of the electrons is large compared to their wavelength, the effect is small and vice versa.

From equations (11.6) and (11.9) we can put the temperature-dependent part of $\Delta\sigma$ very simply by writing $\sigma_0 = (e^2/3\pi^2\hbar)k_F^2 l$ (which follows from equation (3.35) with $v_F\tau = l$ and $S = 4\pi k_F^2$):

$$\Delta\sigma(T) = (e^2/2\pi^2\hbar)(D\tau_{in})^{-1/2} \qquad (11.10)$$

Let me recapitulate this important argument. In materials in which the conduction electrons have a very short mean free path (causing a high resistivity) interference between scattered waves may occur. If the electron waves maintain phase coherence along their path, closed loop paths offer the electron two paths of equal phase change, namely the closed path executed in opposite senses. The two waves thus return to their starting point exactly in phase and reinforce each other, thereby doubling the probability of this path in comparison with classical expectations and in comparison with other open paths where interference effects average to zero. There is thus an enhanced probability of an electron returning to its starting point and this is referred to as weak localisation. It leads to an enhanced resistance at low temperatures but this additional resistance is destroyed as the temperature is raised because of dephasing processes, such as inelastic scattering by phonons. Classical diffusion theory suggests that the temperature dependence of this additional resistance will be governed by the temperature dependence of the term $(\tau_{in})^{-1/2}$ where τ_{in} is the dephasing or inelastic scattering time[1,2].

11.5 The temperature dependence of resistivity

The results embodied in equations (11.9) and (11.10) enable us to understand many features of the behaviour of the conductivity of metallic glasses and other high-resistivity metals. Let me emphasise that the only conditions that are needed to bring about these weak localisation effects are essentially (a) that the mean free path of the conduction electrons should be comparable with their wavelength so that $k_F l$ is small and (b) that any scattering that can randomise the phase of the electron wavefunction is small. In practice, this means that the temperature must be low enough to limit inelastic scattering. These conditions can be met in crystalline as well as in glassy alloys, in non-transition as well as transition metal alloys. In other words we can expect to see the possibility of the resistivity of metals or alloys falling with rising temperature whenever they have sufficiently high resistivity. In qualitative terms this offers a natural explanation of the Mooij correlation.

To determine how the resistivity of a metallic glass depends on temperature, insofar as it is due to weak localisation, we must know how the dephasing of the wavefunctions depends on temperature; often this means knowing how the inelastic scattering of conduction electrons depends on temperature. The most probable source of such scattering is the phonons and we shall assume here that they are responsible for the dephasing. In the next section we shall look more carefully at how coherence is sustained and how it is destroyed.

We have already discussed the scattering of electrons by phonons in a disordered alloy in Chapter 6, where we saw that scattering by the disordered ions could be elastic (without phonons) or inelastic. The probability of inelastic processes with the generation of a phonon increases as T^2 at low temperatures and as T at high temperatures ($T > \theta_D/3$ or so).

The scattering of electrons in a disordered metal by spontaneously generated phonons (that is, those phonons that are not generated by the scattering of electrons by the disordered ions) generally follows this same temperature dependence (as we saw in Chapter 6).

We see therefore that $1/\tau_{in}$ induced by ionic motion varies as T^2 at low temperatures (T much lower than θ_D) and as T at high temperatures. Thus is follows from equation (11.10) that if the temperature-dependent part of the conductivity is determined by this mechanism it must vary as $(\tau_{in})^{-1/2}$, i.e. as T at low temperatures and as $T^{1/2}$ at high temperatures.

If therefore we plot σ versus T at low temperatures it should, if weak localisation dominates the temperature variation, vary linearly with T; by extrapolation to $T = 0$ we can determine σ_0.

Figure 11.7 is a log–log plot of $\Delta\sigma$ against T for $Cu_{50}Zr_{50}$ and $Cu_{50}Hf_{50}$; it shows that $\Delta\sigma$ does indeed vary as T at low temperatures, although this does not persist at still lower temperatures. Other effects come into play and produce a $T^{1/2}$ dependence, as we see in the figure and as we shall discuss in Chapters 12 and 13.

11.6 Coherence and processes that destroy it

11.6.1 Coherence

The scattering processes that leave the electron wavefunction coherent are those which leave all the ions in their original state. This must leave the state of the individual ions unchanged (e.g. their magnetic spin state) and also leave their *collective* state unchanged. Thus the scattering must be a phononless process, analogous to the emission of a γ-ray in the Mössbauer effect. The measure of how the probability of such phononless

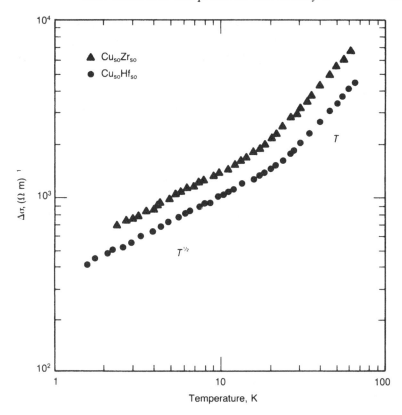

Fig. 11.7 Log–log plot of $\Delta\sigma$ versus temperature in $Cu_{50}Zr_{50}$ and $Cu_{50}Hf_{50}$ metallic glasses to illustrate the T dependence at low temperatures and the $T^{1/2}$ dependence at still lower temperatures. (After Howson and Grieg 1984.)

transitions changes with temperature is the Debye–Waller factor W. The probability of elastic (phononless) scattering, represented by $1/\tau_0$, is reduced by the factor $\exp(-2W)$ which is temperature dependent. The inelastic part of the scattering varies as $1 - \exp(-2W)$, which since $W \ll 1$ can be written simply as $2W$. At low temperatures, in the limit that the glass can be treated as a continuum, W varies as T^2 and at high temperatures, where classical statistics hold, W varies as T. In between, the temperature-dependence depends on the details of the frequency spectrum (normal mode spectrum) of the glass.

Clearly, if the scattering of the electrons by the ionic disorder is to be coherent, there have to be enough such processes to make possible the necessary closed paths for the electrons.

Even at absolute zero, the Debye–Waller factor does not vanish. This is because, as we saw earlier, the zero-point motion of the ions can cause physical effects. Here it causes the ions to have positions that are blurred and in fact their scattering potential is modified by the Debye–Waller factor $W(0)$ at absolute zero. This does not dephase the electron wavefunction but it alters the form factor of the ions. On the Debye model:

$$W(0) = (3\hbar^2 K^2)/8Mk_B\theta_D \qquad (11.11)$$

Here K is the magnitude of the scattering vector of the electron, M the mass of the ions and θ_D the Debye temperature. In fact the Debye–Waller factor as given by the Debye model is not very reliable since the Debye spectrum gives too much prominence to the higher normal mode frequencies compared to those in a real glass.

11.6.2 Processes that destroy coherence

The most obvious such processes are inelastic scattering processes in which the energy state of the electron is altered. Of these the commonest are likely to be scattering by phonons as we have already discussed.

There are, of course, other sources of inelastic scattering and, as we saw, other power laws for phonon scattering. The other sources include electron–electron scattering (see Appendix A6), scattering by ions that carry a magnetic moment (see section 11.9.4) and so on. Moreover, it is possible for the energy of the electron to be changed during scattering without destroying phase coherence: for example, a very low-frequency phonon can compress the metal locally, thereby causing an electron to change direction but because the process is slow the electron wavefunction remains coherent throughout the scattering process. In this case the phonon ultimately causes a change of momentum without a change in energy; such processes are not usually considered in conventional theory because they are second order processes and likely to be of low probability. In general we can take it that phonon scattering destroys the phase coherence of the electron wavefunction but we must bear in mind the complexity and subtlety of the notion of phase coherence; whereas it is often convenient to identify the phase coherence time with the inelastic scattering time this is not always so.

We now turn to the effects of a magnetic field on weak localisation and then to the remarkable consequences of spin–orbit scattering.

11.7 Magnetic field dependence of resistivity

Our discussion of the Aharonov–Bohm effect leads us naturally to expect that the quantum interference effects just described will be altered by a magnetic field. As we saw earlier, the relative phase of the two counter-propagating waves is altered by $2e\phi/\hbar$ if magnetic flux ϕ passes through a closed path executed by an electron. In a high-resistance specimen there are many such orbits of different dimensions and orientations with respect to the applied magnetic field and indeed there is no reason for them to be planar orbits.

To make an estimate of the scale of the effect, consider first a planar closed path normal to the field. We suppose that all the scattering is elastic and that the path length is L with area of order L^2. (The precise coefficient here will depend on the shape of the closed path.) If now a magnetic field B is applied, the flux through the orbit is BL^2 and the consequent change in phase between the two partial waves is $2(e/\hbar)BL^2$. If the motion of the electron along this path is diffusive, the time taken to execute the path is given by $L^2 \simeq Dt$ where as before D is the electron diffusion coefficient. Thus we can think of the dephasing of the partial waves as progressing in time at a rate determined by $(e/\hbar)BDt$, which is independent of the size of the path; this of course applies only to paths of the same shape. Since there are paths of different shapes (not necessarily planar) and of different inclinations to the field direction, the rate of dephasing will vary according to the flux through the particular orbit, i.e. according to its area projected normal to the B-field. Thus, although for a particular orbit the phase change is periodic in the strength of the B-field and will return to the in-phase condition at the appropriate value of B, this will not be true for the whole diverse range of orbits. As the field increases from zero all partial waves are at first dephased; at higher fields the fall in resistivity will be moderated as some orbits approach phase matching again and so partly offset those which are still moving out of phase. The upshot is a general randomising of the phases which completely destroys the constructive interference that gave rise to the enhanced resistance. Thus the magnetic field gradually reduces this resistance and gives rise to the comparatively rare phenomenon, a negative magnetoresistance.

The orbits that contribute most to the dephasing are those which for a given value of L have the greatest projected area normal to B. There is thus a characteristic time-scale after which the constructive interference in such orbits is effectively destroyed. From the argument above, the dephasing at time t is of order $(e/\hbar)BDt$ so that when this reaches the value π the partial waves from these important closed paths will be in antiphase and their contribution to the enhanced scattering will be destroyed. We assume that in this time the range of variation of the phase from other orbits is sufficient to kill any increase in resistivity from the periodic variation of particular groups of orbits.

Let us call this characteristic time τ_B since it refers to a specific value of B; it is defined by the relation:

$$\pi \simeq (e/\hbar)BD\tau_B \quad \text{or} \quad \tau_B \simeq \hbar/eBD$$

This derivation merely suggests the form of the answer and gives some insight into the physics of the process. The correct definition of τ_B is:

$$\tau_B = \hbar/4eBD \qquad (11.12)$$

For a typical metallic glass of the kind we are discussing (with a value of D about $5 \times 10^{-5} \, \mathrm{m^2 \, s^{-1}}$) placed in a magnetic field of 1 T, this has a value of about 10^{-12}–10^{-13} s.

To find out what the magnetic field does to the enhanced resistance, we must now find the probability of generating closed loop paths in this time interval. To do this we must integrate equation (11.5) from τ_0 to τ_B and so get:

$$p(\tau_B) \simeq \lambda^2 v_F [(\tau_0)^{-1/2} - (\tau_B)^{-1/2}]/(4\pi D)^{3/2} \qquad (11.13)$$

This is proportional to the enhanced resistivity in the presence of a magnetic field B. The second term on the right-hand side is the field-dependent part and from equation (11.12) is seen to be proportional to $-B^{1/2}$, i.e. the magnetoresistance at high fields is negative, proportional to the square root of the field and is independent of temperature and field direction. This is a most striking result and quite unlike the behaviour we previously attributed to the magnetoresistance of metals or alloys.

We can estimate the size of this magnetoresistance at low temperatures in the following way. The first term in equation (11.13) is proportional to the enhanced part of the resistivity in zero field whereas the second is proportional to the change due to the field. The ratio of the second term to the first is $(\tau_0/\tau_B)^{1/2}$. For an alloy in a field of 1 T with τ_0 of about 10^{-16} s and $D = 5 \times 10^{-5} \, \mathrm{m^2 \, s^{-1}}$ this ratio is about 10^{-2}. Since at low temperatures the enhanced part of the resistivity is typically about 1 % of the total, the magnetoresistance at 1 T should be of order 10^{-4}. This is enormously larger than any realistic value for the Kohler magnetoresistance, which in such a high-resistivity glassy alloy would be almost immeasurably small. Thus the magnitude and the unusual field dependence ($B^{1/2}$) are clear and striking predictions.

We can rewrite equation (11.13) in terms of the conductivity as follows. We concentrate on the field-dependent term, putting $\tau_B = \hbar/4eBD$ and recognising that $p(\tau_B) = \Delta\rho(B)/\rho_0$. If therefore we multiply both sides by $\sigma_0 = e^2 k_F l/3\pi^2\hbar$, we obtain:

$$\Delta\sigma(B) = \text{constant} \times (e^2/2\pi^2\hbar)(eB/\hbar)^{1/2} \qquad (11.14)$$

where we have also used $D = v_F l/3$ and $\lambda = 2\pi/k_F$. The astonishing feature of this result is that the change in conductivity for a given field is independent of temperature or the metal!

In equation (11.14) the factor $e^2/2\pi^2\hbar$ has the dimension of conductance and has the value of about $(80\text{k}\Omega)^{-1} \simeq 10^{-5}\,\Omega^{-1}$. It recurs in many of the results which follow, just as does the combination $(eB/\hbar)$, which has the dimensions of reciprocal area or reciprocal length squared (L^{-2}). At 1 T, the value of $(eB/\hbar)^{1/2}$ is about $(1/4 \times 10^{-7})\,\text{m}^{-1}$.

Thus the magnitude of $\Delta\sigma$ in equation (11.14) in a field of 1 T, calculated with the correct constant of proportionality, is $290\,(\Omega\text{m})^{-1}$. In making measurements, however, one is concerned with relative changes of conductivity and so, if we take $500\,\mu\Omega\,\text{cm}$ for the resistivity of a metallic alloy, the relative change in a field of 1 T is about 10^{-3}. For higher conductivity metals the effect is correspondingly smaller and may well be masked by other changes. Another way of looking at this result is through the relationship $\Delta\sigma = -\Delta\rho/\rho^2$, which implies that since $\Delta\sigma$ for a given field is fixed, the magnetoresistance $\Delta\rho/\rho \propto \rho$, i.e. the higher the resistivity, the bigger the magnetoresistance from this source.

11.7.1 Magnetoresistance at low fields

This $B^{1/2}$ field-dependence does not persist down to low fields for the following reason. As we saw in the experiment of the hollow cylinder, the effect on a given electron orbit is to produce an oscillatory variation in the resistance as the phase difference between the two partial waves increases with the flux. In the normal geometry and moderate fields, this is not seen because the different orbits all have different periodicities which ultimately randomise the phases. But at very low fields, all the orbits have their phases changed in concert and so all contribute together to dephase the partial waves and lower the resistance. Suppose that two waves of the same amplitude A combine to produce an intensity I_0 when in phase. If they now differ in phase by δ, the new combined amplitude is $2A \cos(\delta/2)$ and the intensity will then be reduced to:

$$I_0 \cos^2(\delta/2) \simeq I_0(1 - \delta^2/4) \qquad (11.15)$$

if δ is small. Here the phase changes are small at low fields and, with $\delta = (e/\hbar)BL^2$, all are proportional to B. Each closed orbit will contribute such terms and so the field-dependent terms add up to give a negative term proportional to B^2, which determines the low-field magnetoresistance. The other factor is L^4, which is limited by the possible path lengths. At a given temperature $L^2 \propto D\tau_{\mathrm{in}}$ and so L^4 contributes a factor $(D\tau_{\mathrm{in}})^2$ in the numerator. On the other hand I_0 has a factor $(D\tau_{\mathrm{in}})^{1/2}$ in the denominator (see equation (11.10)) so that the field-dependent term varies as $-(eB/\hbar)^2(D\tau_{\mathrm{in}})^{3/2}$. This is confirmed by a full calculation.

The dependence on τ_{in} shows that unlike the high-field effect the low-field magnetoresistance depends on temperature and thus measurements of the B^2 magnetoresistance enable one to find τ_{in} and its temperature dependence. Such measurements have been made on a number of metallic glasses at low temperatures (see Chapter 16) and many of them show that τ_{in} varies as T^{-2}.

As the temperature falls, thereby reducing τ_{in}, the $B^{1/2}$ dependence persists to lower and lower fields. This is because larger and larger orbits come into play and in these a small change of field produces a big change of flux. Consequently the disparity in phase between the largest and smallest orbits increases and the B^2 region in which all the phases change in harmony gets smaller. At absolute zero the magnetoresistance varies as $B^{1/2}$ over the full range of fields.

Conversely, as the temperature rises, the B^2 dependence persists to higher and higher fields. Indeed the weak localisation contribution to the magnetoresistance of amorphous $Cu_{65}Ti_{35}$ has been observed up to temperatures of 85 K and shown to vary as B^2 up to a field of 12 T at this temperature (Lindqvist 1992).

The full expression for the change in conductivity at low fields due to weak localisation, in the absence of spin–orbit effects (see below), is given in Appendix A1.

So far we have assumed that there is only one type of charge carrier in the alloy, although we know that when one or more component is a transition metal we can expect to have s- and d-electrons present. The theory outlined above can still be applied if the parameters involved are properly interpreted. This is explained in Appendix A2.

These predictions for the magnetoresistance, that it should be negative, vary as B^2 at low fields and vary as $B^{1/2}$ at high fields, cannot yet be fully compared with experiment because other effects that we have not yet discussed occur simultaneously, in particular, the influence on these quantum interference effects of the electron spin and the consequent importance of spin–orbit coupling in the scattering. We now turn to these matters.

11.8 The influence of electron spin

We have been looking at the special properties of closed electron paths and their influence on the resistivity of metallic glasses without considering the effect of electron spin. We must now remedy this.

The two partial electron waves that propagate in opposite senses round the closed path can either have the same or have opposite spin directions and these two combinations have opposite effects when the partial waves interfere. If the two partial waves have the same spin they recombine after counter-propagation to enhance the probability above its classical value. If however they carry opposite spins they recombine to reduce the classical probability.

There is a further important point. We must recognise that if we have a pair of electrons they can form four independent spin functions, which we can represent by (↑↑), (↑↓), (↓↑) and (↓↓). These in turn can be combined into orthonormal combinations with the total eigenvalues projected in the field direction as indicated in Table 11.2.

The first three form the triplet combination, often written with total spin $j = 1$ and projections in the magnetic field direction $m = +1$, 0, -1; the last term forms the singlet combination with $j = 0$ and $m = 0$. These combinations will also be important to us when we discuss the interaction between electrons.

If therefore the two partial waves have opposite spins there is only a single state corresponding to this but if the two partial waves have the same spin direction there are three such states. The singlet and triplet states just described are exactly analogous to those familiar from the spectroscopic states of electrons in atoms. Incidentally, and the significance of this will be clear later, we are assuming so far that there is no spin–orbit coupling involved in the scattering.

We wish now to write down the probability of an electron executing a closed path as in equation (11.3) but taking account of electron spin. Since the classical terms are unchanged, let us concentrate on the inter-

Table 11.2 *Spin states of electron pairs*

Combination	Eigenvalue	m
$(\uparrow\uparrow)$	h	$+1$
$2^{-1/2}[(\uparrow\downarrow) + (\downarrow\uparrow)]$	0	0
$(\downarrow\downarrow)$	$-h$	-1
$2^{-1/2}[(\uparrow\downarrow) - (\downarrow\uparrow)]$	0	0

ference terms only. We denote by ψ_{00} the singlet state in which the spins are in opposition and by ψ_{11}, ψ_{10} and ψ_{1-1} the three components of the triplet state. Then the interference term I is given by:

$$I = (1/2)\{|\psi_{11}|^2 + |\psi_{10}|^2 + |\psi_{1-1}|^2 - |\psi_{00}|^2\} \qquad (11.16)$$

In the absence of spin–orbit effects the first three terms are each equal (numerically) to the fourth so that $I = |\psi_{00}|^2$; i.e. the interference term is positive and just equal to the classical value as we saw earlier when we ignored the spin variable. In the presence of spin–orbit effects this is no longer generally true.

11.9 Spin–orbit scattering (*see also section 9.4*)

Within an atom the spin of an electron may couple to its orbital motion. Formally the interaction looks like the interaction of two magnetic dipoles and depends on the angle between the orbital angular momentum and the spin. However its origin is not magnetic but electrostatic. Its strength rises sharply with the nuclear charge and so the effect is mainly associated with the heavy elements. As we saw in section 9.4 on skew scattering, conduction electrons are also influenced by spin–orbit coupling when they penetrate inside the core of an ion, the orbital motion of the electron being simply its angular motion inside the ion. The spin–orbit scattering then depends on the product:

$$\sigma_z \cdot \mathbf{k} \times \mathbf{k}' \qquad (11.17)$$

where $\mathbf{k}$ and $\mathbf{k}'$ are the wave vector of the electron before and after scattering and σ_z is the unit vector in the magnetic field direction, $+$ for spin up and $-$ for spin down. Since in this expression the dot and cross can be interchanged, we can think of the scattering as dependent on the vector product between the spin direction vector and the wave vector of the incident electron. Thus, as we shall now see, the scattering has different effects according to the spin states of the partial waves executing a closed path.

If the two waves have opposite spins then, since they perform the path in opposite senses and their velocities are also opposite, the angle between the spin and velocity is the same at corresponding scattering events. Thus the scattering processes are the same for the two waves, though performed in inverse order, and the two arrive back at their starting point in phase. If the two waves have the same spin direction, however, the two partial waves now behave quite differently. Their spin directions are initially the same but their velocities round the path are reversed so that induced changes of spin direction are of opposite sign. At each scattering event the initial conditions are different and consequently so are the final ones.

When the partial waves return to their starting point their phases tend to be randomised and so to cancel each other. The time this takes to occur depends on the strength of the spin–orbit coupling and on the concentration of such scattering centres; it is usually designated τ_{so} with a typical value for the 50–50 alloys in the middle range of atomic numbers of 10^{-12}–10^{-13} s independent of temperature. It can of course be deliberately shortened by doping the specimen with a heavy metal such as gold.

11.9.1 Spin–orbit scattering and temperature-dependence

To calculate the quantum interference effect in the presence of spin–orbit scattering, we recognise that the singlet contribution to σ is positive and unchanged by spin–orbit effects, whereas the triplet contribution is negative and decays exponentially with a time constant τ_{so}. We can thus rewrite the integrated form of equation (11.5) as:

$$\Delta\sigma/\sigma \simeq - \int [\lambda^2 v_F dt/Dt^{3/2}]\{(3/2)\exp(-t/\tau_{so}) - 1/2\} \qquad (11.18)$$

In this all the terms are just as in equation (11.5) except that an additional factor in the curly brackets has been added. In the curly brackets the exponential factor describes the decay of the triplet term from spin–orbit scattering. This term has a threefold weighting compared to the constant term $-1/2$, which is the singlet contribution. The factor of $1/2$ in both terms comes from the same factor in equation (11.16).

If τ_{so} is long compared to the inelastic scattering time τ_{in}, the triplet term is hardly altered by the exponential decay and the curly bracket has the value unity throughout the time of integration. Then the spin–orbit coupling has almost no effect. If, however, τ_{so} is short compared to τ_{in}, the triplet contribution is quickly wiped out; only the singlet term survives unchanged and is negative. The quantum interference term is now *negative* instead of positive and the forward (not the backward) scattering is enhanced. This new phenomenon induced by the strong spin–orbit scattering is referred to as anti-localisation and all our previous results are turned on their head. What before was positive is now negative and vice versa (see Table 11.3).

The qualitative consequences of spin–orbit scattering for the temperature dependence of the conductivity can be made clear by means of Figure 11.8, where the changes induced by weak localisation are plotted with reference to Boltzmann theory. We shall often find that this brings out clearly the physics of the situation.

As we have just seen, there are two contributions to $\Delta\sigma$, one from partial waves with parallel spins $\Delta\sigma(\uparrow\uparrow)$ and the other $\Delta\sigma(\uparrow\downarrow)$ from those with antiparallel spins; they are of opposite signs and, in the absence of spin–orbit scattering, $|\Delta\sigma(\uparrow\uparrow)|$ is three times as big as $|\Delta\sigma(\uparrow\downarrow)|$. Let us suppose for simplicity that the dephasing probability is due to inelastic scattering by phonons with $1/\tau_{in}$ varying as T^2. This means that both $\Delta\sigma(\uparrow\uparrow)$ and $\Delta\sigma(\uparrow\downarrow)$ change linearly with T as the temperature rises from absolute zero; this is illustrated in Figure 11.8(a). The $\Delta\sigma$'s are measured from the Boltzmann value of the conductivity, which we assume, again for simplicity, is so dominated by elastic scattering from the disorder that it is temperature independent. The total change in conductivity, which is the algebraic sum of $\Delta\sigma(\uparrow\uparrow)$ and $\Delta\sigma(\uparrow\downarrow)$, is shown in Figure 11.8(b) by the lowest curve. Now we turn on the spin–orbit scattering, which alters only $\Delta\sigma(\uparrow\uparrow)$. At absolute zero, the effect is to reduce the value of $\Delta\sigma(\uparrow\uparrow)$; at higher temperatures $\Delta\sigma(\uparrow\uparrow)$ stays constant at this absolute zero value as long as spin–orbit scattering is dominant (i.e. $1/\tau_{so} \gg 1/\tau_{in}$). When, however, the inelastic phonon processes become appreciable with rising temperature $\Delta\sigma(\uparrow\uparrow)$ begins to change with temperature and, when these phonon processes dominate the scattering $(1/\tau_{in} \gg 1/\tau_{so})$, $\Delta\sigma(\uparrow\uparrow)$ behaves as if there was no spin–orbit scattering, i.e. it joins the original curve that represents zero spin–orbit scattering. This is illustrated for an intermediate value of the spin–orbit scattering in Figure 11.8(a) and the consequence for $\Delta\sigma$(total) in Figure 11.8(b).

Each increase in the spin–orbit scattering reduces the absolute zero value of $\Delta\sigma(\uparrow\uparrow)$ and its subsequent low-temperature contribution; in the extreme limit, spin–orbit scattering destroys $\Delta\sigma(\uparrow\uparrow)$ altogether.

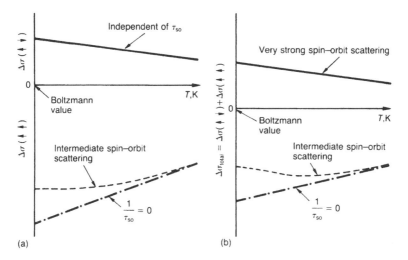

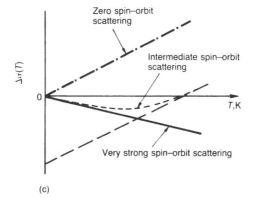

Fig. 11.8 Temperature dependence of the conductivity due to weak localisation in the presence of spin–orbit scattering (schematic). (a) Parallel and antiparallel spin contributions shown separately with the Boltzmann value chosen as zero. The scattering is assumed to be from phonons at low temperatures with $1/\tau_{in}$ varying as T^2. The upper curve $\Delta\sigma(\uparrow\downarrow)$ is independent of spin–orbit scattering. The lower curve $\Delta\sigma(\uparrow\uparrow)$ is modified by intermediate spin–orbit scattering as shown; this contribution can be destroyed entirely if, in the range of measurement, $1/\tau_{so} \gg 1/\tau_{in}$. (b) The total conductivity change, being the sum of $\Delta\sigma(\uparrow\uparrow)$ and $\Delta\tau(\uparrow\downarrow)$, under conditions of very strong (top curve), intermediate (dashed curve) and zero (bottom curve) spin–orbit scattering. Notice that the intermediate spin–orbit scattering curve starts parallel to the very strong spin–orbit scattering line and ultimately joins that of zero spin–orbit scattering. (c) The total conductivity change with a new origin for which $\Delta\sigma(T)$ is zero at $T = 0$. Notice that the curve for intermediate spin–orbit scattering starts off along the very strong spin–orbit scattering line but becomes parallel to that of zero spin–orbit scattering at high enough temperatures.

Table 11.3 *Some effects of strong elastic scattering through weak localisation*

	Effect of weak localisation	
Property	Without spin–orbit scattering[a]	With strong spin–orbit scattering
Conductivity	$\Delta\sigma$ increasing $\Delta\sigma \propto T$ (phonons at low temperature[b])	$\Delta\sigma$ decreasing $\Delta\sigma \propto T$ (phonons at low temperature[b])
Magnetoresistance	$\Delta\rho(B)/\rho$ negative $\propto B^2$ at low fields $\propto B^{1/2}$ at high fields	$\Delta\rho(B)/\rho$ positive at low fields negative at very high fields

[a] There is always some spin–orbit scattering, so this is an idealisation.
[b] The temperature dependence caused by phonon scattering may differ from T.

Then the weak-localisation contribution to the conductivity comes entirely from $\Delta(\uparrow\downarrow)$. This is shown by the topmost curve in Figure 11.8(b).

Now of course the Boltzmann zero is not accessible to experiment so that to see what experimental measurements would show, we must plot $\Delta\sigma$(total) versus temperature from a common origin as in Figure 11.8(c). Notice that all the curves start from absolute zero with the same positive slope, that of $\Delta\sigma(\uparrow\downarrow)$, because when there is any spin–orbit scattering, no matter how small, $\Delta\sigma(\uparrow\uparrow)$ sets out from absolute zero with zero slope, leaving the unchanged $\Delta\sigma(\uparrow\downarrow)$ to determine that of the total. A further point is that, once phonon scattering dominates over spin–orbit scattering, the curve of $\Delta\sigma$(total) is parallel to that with zero spin–orbit scattering, but displaced by the shift of origin.

These very striking effects of spin–orbit scattering can be derived essentially from equation (11.18) by recognising that, whereas the singlet contribution is unchanged, the probability of scattering (per unit time) in the presence of both inelastic and spin–orbit scattering is given for the triplet terms by:

$$1/\tau_{\text{total}} = 1/\tau_{\text{in}} + a/\tau_{\text{so}} \qquad (11.19)$$

where a is a weighting factor to take account of spin degeneracy and the definition of τ_{so}.

Thus for the singlet term we integrate the classical probability as before from τ_0 to τ_{in} because the spin–orbit scattering has no effect on this term. For the triplet terms, however, the upper limit of integration is now τ_{total} in equation (11.19). The resulting expression for the temperature-dependent part of the conductivity is given by:

$$\Delta\sigma(T) \simeq (e^2/4\pi^2)D^{-1/2}[3(1/\tau_{in} + a/\tau_{so})^{1/2} - 3(a/\tau_{so})^{1/2} - (1/\tau_{in})^{1/2}]$$

(11.20)

The constant term has been chosen to ensure that $\Delta\sigma$ starts from zero when $T = 0$. (See also Appendix A1 for more details.)

11.9.2 Spin–orbit scattering and magnetoresistance

There is a further important consequence of spin–orbit scattering. If a magnetic field is applied, we know that it causes the partial waves to get out of phase in a time τ_B which decreases as the field gets stronger. If this time becomes shorter than the spin–orbit scattering time, spin–orbit scattering has insufficient time to operate and spin–orbit effects disappear.

As with the temperature dependence, the magnetic field dependence can be illustrated qualitatively by a suitable diagram. First let us consider the effect of a magnetic field at absolute zero; we see in Figure 11.9 the separate contributions $\Delta\sigma(\uparrow\uparrow)$ and $\Delta\sigma(\uparrow\downarrow)$ measured from the Boltzmann value as zero. As before they are of opposite signs and if there is no spin–orbit scattering $|\Delta\sigma(\uparrow\uparrow)| = 3|\Delta\sigma(\uparrow\downarrow)|$. As the magnetic field is increased, the two contributions vary as $B^{1/2}$ and so, if we plot the changes against $B^{1/2}$, the changes are linear as shown in Figure 11.9(a). The total change is the algebraic sum of the two parts as shown in Figure 11.9(b).

We now turn on the spin–orbit scattering, which is a random dephasing mechanism that reduces the contribution from $\Delta\sigma(\uparrow\uparrow)$ but leaves $\Delta\sigma(\uparrow\downarrow)$ unchanged; this is shown for one intermediate value of the spin–orbit scattering in Figure 11.9(a). The effect of spin–orbit scattering on $\Delta\sigma(\uparrow\uparrow)$ is formally the same as inelastic scattering in that at low fields (see section 11.7.1) it causes $\Delta\sigma(\uparrow\uparrow)$ to increase as B^2. Because at $B = 0$ the slope due to $\Delta\sigma(\uparrow\uparrow)$ is thus zero, the initial slope of $\Delta\sigma(total)$ is determined by that of $\Delta\sigma(\uparrow\downarrow)$, i.e. as $B^{1/2}$ as shown in Figure 11.9(b). Finally at high fields where $1/\tau_B \gg 1/\tau_{so}$, $\Delta\sigma(\uparrow\uparrow)$ is unaffected by spin–orbit scattering and resumes the value it would have in its absence. This is shown in Figure 11.9(a) and the resulting values of $\Delta\sigma(total)$ in

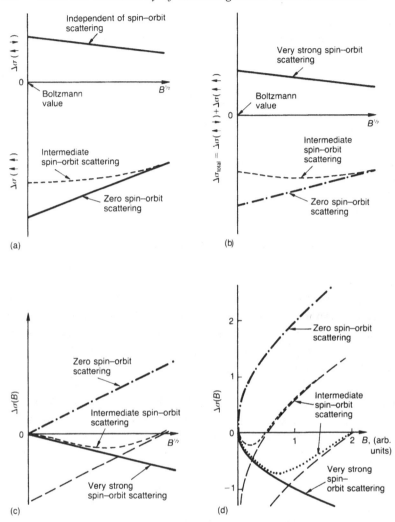

Fig. 11.9 The effect of spin–orbit scattering on the magnetoconductivity due to weak localisation (schematic). (a) Parallel and antiparallel spin contributions at $T = 0$ plotted against $B^{1/2}$ with reference to the Boltzmann value as zero. $\Delta\sigma(\uparrow\downarrow)$ is unchanged by spin–orbit scattering but $|\Delta\sigma(\uparrow\uparrow)|$ is reduced as shown by the dashed line, which starts at low fields as B^2; with sufficiently strong spin–orbit scattering $\Delta\sigma(\uparrow\uparrow)$ can be made to vanish within the range of fields used. (Only at $T = 0$ and with no spin–orbit scattering does the $B^{1/2}$ behaviour persist down to $B = 0$; at higher temperatures the curves start as B^2.) (b) The total conductivity change, being the sum of $\Delta\sigma(\uparrow\uparrow)$ and $\Delta\sigma(\uparrow\downarrow)$, under very strong (top curve), intermediate (dashed curve) and zero spin–orbit scattering at $T = 0$. These curves are analogous to those of Figure (11.8). (c) The same curves as in (b) but now referred to a common origin. (d) The same curves as in (c) but plotted against B instead of $B^{1/2}$.

Figure 11.9(b). Figure 11.9(c) shows $\Delta\sigma(B)$ versus B where all the curves have a common origin to correspond to what is measured experimentally. Figure 11.9(d) shows schematically what Figure 11.9(c) would look like if the changes were plotted against B instead of $B^{1/2}$.

All this refers to the magnetoresistance at absolute zero. At higher temperatures, phonon scattering changes *both* $\Delta\sigma(\uparrow\uparrow)$ and $\Delta\sigma(\uparrow\downarrow)$, reducing them proportionately at zero field and tending to keep them constant at low magnetic fields. Spin–orbit scattering alters only $\Delta\sigma(\uparrow\uparrow)$ and so, as the field is increased, the total conductivity rises slowly (as B^2 at first) until when τ_B becomes comparable with τ_{in} the conductivity begins to recover the value it would have in the same field at absolute zero. At absolute zero, moreover, spin–orbit scattering always produces regions of positive magnetoresistance at low fields. Phonon scattering reduces these and ultimately eliminates them completely. This occurs when $1/\tau_{in}$ becomes so large compared to $1/\tau_{so}$ that it suppresses any change in $\Delta\sigma(\uparrow\downarrow)$ up to fields large enough to 'kill' the spin–orbit scattering. Appendix A1 gives expressions for the magnetoconductivity in the presence of spin–orbit scattering.

11.9.3 Experimental evidence for spin–orbit scattering

Table 11.3 sums up the changes in resistivity due to quantum interference by contrasting the behaviour with and without strong spin–orbit scattering. These are striking predictions. The following examples where the magnetoresistance is due to quantum interference demonstrate their validity.

Figure 11.10(a) shows the magnetoresistance of amorphous $Mg_{80}Cu_{20}$ at 4.2 K; it is negative because spin–orbit effects are small in this alloy. Figure 11.10(b) shows the magnetoresistance of amorphous $Cu_{50}Lu_{50}$; it is positive because here spin–orbit effects are large. In Figure 11.10(c) the alloy is amorphous $Cu_{50}Y_{50}$ in which spin–orbit effects are of moderate strength. Consequently at low fields where the magnetic scattering time τ_B is long compared to τ_{so}, the spin–orbit scattering has time to destroy the coherence of the triplet state waves and the magnetoresistance is positive. At higher fields τ_B diminishes (it varies as $1/B$) and becomes too short to allow spin–orbit effects to be fully effective; the magnetoresistance then ceases to grow, begins to fall and would ultimately become negative at high enough fields. These results are discussed more fully in Chapter 16.

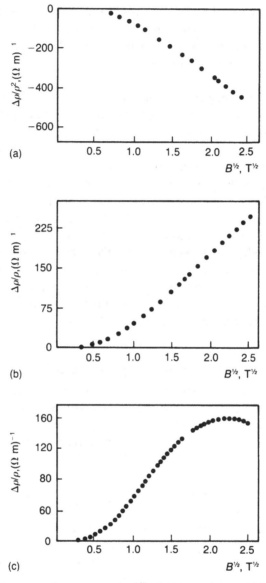

Fig. 11.10 Magnetoresistance versus $B^{1/2}$ of (a) $Mg_{80}Cu_{20}$, which has only small spin–orbit scattering, so the magnetoresistance is negative at all appreciable fields; (b) $Cu_{50}Lu_{50}$, in which the spin–orbit scattering is strong enough to make the magnetoresistance positive throughout the field range; and (c) $Cu_{50}Y_{50}$, where the spin–orbit scattering is of moderate strength so that the magnetoresistance starts positive, reaches a maximum and would ultimately go negative at high enough fields. (The ordinates here are $\Delta\rho/\rho^2 = -\Delta\sigma$ and so have the dimensions of conductivity.) (After Bieri *et al.* 1986.)

Figure 11.11 shows how adding 3 atomic % gold to a Ca–Al metallic glass changes the magnetoresistance from negative (because spin–orbit effects are small in Ca and Al) to positive because of the large spin–orbit scattering from the gold.

These remarkable results of spin–orbit scattering demonstrate the range and diversity of the effects predicted to follow from weak localisation.

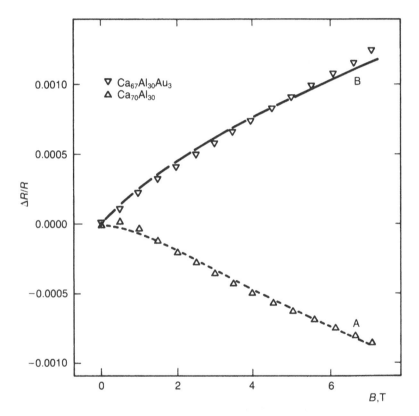

Fig. 11.11 Effect of adding 3% Au to CaAl glass, showing how the sign of the magnetoresistance is reversed. In curve A spin–orbit scattering is negligible: in curve B the alloy has been doped with gold to introduce strong spin–orbit scattering. (After Howson *et al.* 1988.)

11.9.4 Spin–spin or spin–flip scattering

The presence of magnetic impurities which carry a local magnetic moment can cause the partial waves to get out of phase; this comes about in the following way. When the electron is scattered by the magnetic impurity, the spin direction of the incident electron is rotated in the process and, since the two partial waves travel in opposite directions, the rotations of the spin occur in reverse sequence for the one as compared to the other. In three dimensions rotation A followed by rotation B does not in general produce the same result as B followed by A. This means that the final spin directions of the two partial waves are different and their phases become randomised. This is true of both types of partial waves, those with parallel and those with antiparallel spins: both are put out of phase and so both contributions to the back-scattering are reduced. In this respect magnetic impurities produce effects quite different from spin–orbit scattering, which alters only the triplet contribution from the partial waves.

To emphasise the difference between the two types of scattering, we can say that in spin–orbit scattering, the angle between the momentum of the electron and its spin direction is important whereas in spin–flip scattering it is the relative spin directions and sequence of scattering events that matter.

Before we conclude this chapter we will look at a method of deriving the general form of the results of weak localisation from a very different point of view.

11.10 Scaling theory and weak localisation[3]

It is possible to derive some of the results already derived on quantum interference by rather general arguments with very few assumptions. This is achieved by means of scaling theory applied to non-interacting electrons. Such theories are concerned with the response of systems when their properties reach extreme values, and their size but not their physical shape is changed; they demonstrate how the dimensionality of the system alters its response. They can give information about the approach to the metal–insulator transition when the conductance of a system becomes very small; they can also identify the important parameters of the system and the range of validity of certain approximations, in particular perturbation treatments.

We shall be concerned only with the range of validity of weak localisation theories and what scaling theory can tell us about the form of such effects. It is not surprising that the information is not quantitative but it does serve to define the limits of Boltzmann theory.

The electrical conductance G of a system is just the reciprocal of its resistance R and is a convenient quantity to work with. If the conductivity is σ and its resistivity ρ, we have:

$$R = l\rho/A \quad \text{and} \quad G = A/l\rho \tag{11.21}$$

Here l is the length and A the cross-section of the specimen. Clearly if we are in the normal regime where ρ and σ are independent of size and if L is a measure of the linear size of the system, R scales with L^{-1} and G with L in three dimensions. In two dimensions R and G are independent of scale and in one dimension R goes as L and G as L^{-1}.

According to Boltzmann theory we can write:

$$\sigma_0 = (e^2/12\pi^3\hbar)S\lambda \tag{11.22}$$

where S is the area of the Fermi surface and λ is used to denote the mean free path of the electrons averaged over S (λ is used instead of l to avoid confusion with the length of the sample). For a spherical surface we can write $S = 4\pi k_F^2$. This shows that if we wish to define conductance in dimensionless form in three dimensions we must put:

$$G = (\hbar/e^2)\sigma L \tag{11.23}$$

where L as before is a linear measure of the system size.

The quantity commonly used in these discussions is the dimensionless derivative $\beta = \mathrm{d}\log G/\mathrm{d}\log L$. In the regime where σ is independent of size we know from what was said earlier that this derivative is a constant; its value is $+1$ in three dimensions, 0 in two and -1 in one dimension. The scaling theory of non-interacting electrons of Abrahams *et al.* (1979) indicates that there must be departures from this constancy and that a plot of $\mathrm{d}\log G/\mathrm{d}\log L$ versus $\log G$ would have the form shown in Figure 11.12.

Many authors have worked on this problem. Thouless for example has shown how three-dimensional wires take on the characteristics of one dimension at high enough resistance. This means that $\mathrm{d}\log G/\mathrm{d}\log L$ for three dimensions ($d = 3$) must move in the direction of the one-dimensional ($d = 1$) curve in Figure 11.12 as the conductance diminishes. This is indicated in the figure.

This also means that as G decreases (increasing resistance) the logarithmic derivative falls below $+1$ for three-dimensional systems and eventually goes negative. Let us therefore take the simplest representation of this and write:

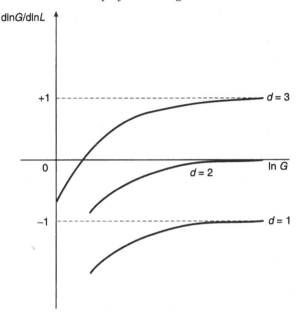

Fig. 11.12 dlog G/dlog L versus the logarithm of the conductance G to show scaling effects (schematic); L is a linear dimension of the sample.

$$d \log G / d \log L = 1 - A/G + \text{higher-order terms in } 1/G \qquad (11.24)$$

Here A is a constant of order unity. We add the condition that $G = G_0$ when $L = L_0$, the minimum value of L at which Boltzmann theory holds i.e. the minimum value for which the derivative β remains essentially $+1$.

If we integrate (11.24) we then get:

$$(G - A)/(G_0 - A) = L/L_0 \qquad (11.25)$$

or $$G = L(G_0 - A)/L_0 + A \qquad (11.26)$$

We now write this in terms of the conductivity with $G = (h/e^2)\sigma L$ and $G_0 = (h/e^2)\sigma_0 L_0$ and get:

$$\sigma = \sigma_0 - A(h/e^2)[(1/L_0) + (1/L)] \qquad (11.27)$$

Finally we set $\sigma - \sigma_0 = \Delta\sigma$ and put $L_0 = \lambda$, the mean free path of the electrons; this must be the value of L at which d$\log G$/d$\log L$ departs from unity because below this the conductivity becomes size-dependent. In the limit of L large compared to L_0 we therefore get:

$$\Delta\sigma/\sigma_0 = -Ah/e^2\sigma_0\lambda \qquad (11.28)$$

We consider the conductivity at absolute zero so that $\lambda = l_e$, the elastic mean free path. We use equation (11.22) with $S = 4\pi k_F^2$ and find:

$$\Delta\sigma/\sigma_0 = -3\pi^2 A/(k_F l_e)^2 \qquad (11.29)$$

This can be compared with equation (11.9), with τ_{in} put equal to infinity, and shows how this rather general argument leads to a correction at absolute zero with the same qualitative features as our more detailed argument.

11.11 Summary

The argument of this chapter has been that in highly disordered metals, of which metallic glasses are excellent examples, the short mean free path of the conduction electrons implies that interference between incident and scattered waves can be important. This feature, which is neglected in Boltzmann theory, gives a special prominence to electron paths that form closed loops because there are then two paths of equal length for the electron wavefunction, namely the closed path executed in opposite senses. If therefore the electron wavefunctions retain phase coherence the two partial waves can reinforce each other after traversing these two paths and double the probability of such paths by comparison with classical expectations. The quantum probability can thus be deduced from the classical theory of diffusion applied to the electron paths.

This enhanced back-scattering, referred to as weak localisation, reduces the conductivity of the metal at absolute zero but as the temperature is raised phonon scattering can destroy the phase coherence and cause the resistance to fall. Likewise a magnetic field applied to the metal can cause the partial waves to get out of phase and so cause the resistance to fall. Thus a negative temperature coefficient of resistance and a negative magnetoresistance are natural consequences of the high resistivity and are indeed found experimentally. Both are unusual features according to Boltzmann theory. Spin–orbit scattering can reverse the sign of these two effects and the theory accounts for these reversals in a natural way. The agreement between theory and experiment in bulk metallic glasses and in, for example, thin crystalline films (not discussed here) is impressive.

12

The interaction effect or Coulomb anomaly

12.1 Introduction

There is a further effect that arises in systems in which there is heavy elastic scattering of the conduction electrons; it shows itself at low temperatures through the unusual temperature and magnetic field dependence of the electrical resistance and since its contribution can be confused with that from weak localisation it is important to describe its consequences before we try to complete the survey of that effect.

The localisation effect described in the last chapter involves single electrons and would exist even if these electrons did not interact with each other. By contrast this new effect, sometimes called the Coulomb anomaly, arises ultimately from the interaction of one electron with another. Hence its rather uninformative alternative name 'the interaction effect', which does however emphasise that it could not occur with non-interacting electrons. The 'enhanced interaction effect' is perhaps a better name[1].

If, in an ordered metal, an electron in a plane wave state of wave vector $\mathbf{k}$ is scattered into state $\mathbf{k}'$, we must have $\mathbf{k}' = \mathbf{k} + \mathbf{q}$ where $\mathbf{q}$ is a Fourier component of the scattering potential. In a disordered metal, however, there is an uncertainty in $\mathbf{k}$ because of the scattering; this uncertainty is of order $1/l$, where l is the relevant mean free path. Thus the above relation will break down if the scattering vector q is less than $1/l$. This suggests that any unusual effects will occur at small q and that our interest will focus on states for which $ql < 1$; this means that the smaller the mean free path involved, the greater the range of q-vectors that can contribute to the effects.

An electron that is repeatedly scattered by the disordered ions can be thought of as diffusing through the material. At large distances and time intervals, this diffusive motion must approximate to classical diffusion and we use this classical picture to describe what happens: the diffusive motion of the electrons alters the time dependence of their motion and hence the energy dependence of their properties; this is reflected in changes to the density of electron states and the conductivity.

The behaviour of the conduction electrons is obviously very complex and much of it contributes nothing of interest. Theory, usually based on the use of Feynman diagrams, tries to pick out those pairs of electrons that behave in special ways that lead to significant contributions to the thermodynamic and transport properties of their host. There are various modes of interaction, of which some involve closed paths like those discussed in weak localisation so it is no accident that the two effects show similarities.

What is important to realise at the outset is that *the fundamental behaviour of the electrons that leads to these changes arises by chance or, to be more precise, from the randomness implicit in the diffusive motion, and not from the Coulomb interaction itself*. The Coulomb interaction is, of course, necessary to make the effects manifest and, in part, determines their size but it does not of itself produce the effects. Finally we note that the diffusive motion of the electrons alters the screening of electron pairs that are widely separated and that this alters the strength of the interaction.

Let us now look at the effect in more detail. In order to make the argument as simple as possible we begin by deriving the effect of the interaction on an equilibrium property, the density of electron states at absolute zero. The Einstein relation connects this to the conductivity and so we shall achieve some insight into this property as well.

12.2 Electron–electron interaction and the density of states

To calculate the density of states for a free-electron gas (section 3.2), we use the fact that the states are labelled by their k-vectors and that these states are uniformly distributed in k-space. Then from the density of states in k-space we can easily derive the density in energy from the E–k relation of the electrons at the Fermi level; this involves dE/dk, which is essentially the electron group velocity. Here we do something quite different: we assume we know the density of states of the non-interacting electrons and then calculate how this is changed by the interaction.

The change in density of states brought about by the interaction is derived by finding out how the energy of a given state, originally E, is altered when the interaction is turned on. Suppose the new energy is E^*, then:

$$E^* = E + S_E \qquad (12.1)$$

where S_E is the so-called self-energy. Clearly, if S_E is independent of E, the energy of all the states is altered by the same amount and the density of states is unchanged. In fact (see Figure 12.1), the change in density of states depends on the rate of change of S_E and is given by:

$$\Delta N(E)/N(0) = -\partial S_E/\partial E \qquad (12.2)$$

To calculate S_E we imagine a single electron at an energy ϵ above the Fermi level and a single vacant state or hole in the Fermi sea (this is thus the so-called 'particle–hole channel'). If the interaction energy between the electron at ϵ and one at energy ϵ' is $s(\epsilon, \epsilon')$, the total self energy is:

$$S_E = \sum s(\epsilon, \epsilon') = \int_{-\infty}^{0} N(\epsilon')s(\epsilon, \epsilon')\mathrm{d}\epsilon' \qquad (12.3)$$

where the sum is over all occupied states below the Fermi level and, in the second part of equation (12.3), the sum has been replaced by an integral with ϵ' going from $-\infty$ to 0 at the Fermi level. Ultimately, however, we are going to differentiate equation (12.3) with respect to ϵ and so if the interaction energy depends, as we assume[2], only on the difference in energy $\hbar\omega = \epsilon - \epsilon'$, we can change the variable from ϵ' to ω. Equation (12.3) then becomes:

$$S_E = -\int_{\epsilon/\hbar}^{\infty} N(\epsilon - \hbar\omega)s(\omega)\hbar\mathrm{d}\omega \qquad (12.4)$$

and the limits on ω are $\epsilon/\hbar$ and $+\infty$. When we differentiate with respect to ϵ, the lower limit, this cancels out the integral, changes the sign again and leaves only the contribution from a state at the Fermi level. The upshot is that the relative change in the density of states at energy ϵ above the Fermi level is given by:

$$\Delta N/N(0) = -s(\epsilon, 0)\,N(0) \qquad (12.5)$$

This is a valuable simplification and reduces our problem to the calculation of the energy of interaction between two electrons, one at ϵ above E_F and the other at E_F. Of course if the new one-electron energies are to be

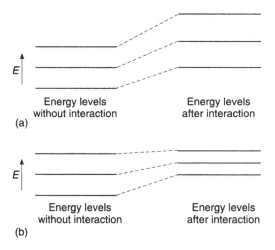

Fig. 12.1 Electron energy levels before and after interaction, with the interaction (a) increasing with energy; (b) decreasing with energy.

meaningful, the self-energy must be small compared to the original energy of the state.

Our problem now is to calculate this interaction energy between two electrons. To do this we have to understand (1) the interaction forces between electrons; (2) the perturbation method of calculating the inter-action energy, which in turn involves wavefunctions that contain Hartree and exchange terms; (3) the special form of the exchange term; and (4) how to take account of the diffusive motion of the electrons in a disor-dered metal. We tackle these in turn.

12.3 Interaction energy in a disordered metal

12.3.1 The Coulomb interaction

The main interaction between a pair of electrons is the Coulomb repul-sion, although as we have already seen there is an attractive force mediated by the phonons. For simplicity we concentrate first on the Coulomb force. A given electron will experience a repulsion from another electron while the remaining mobile electrons and the ions, which together produce overall electrical neutrality, provide screening; the total effect is to reduce the force from one of long-range to one that is exponentially damped. If the two electrons are a long way apart (many interatomic spacings) one electron will be almost unaffected by the force from the other because of this electrical neutrality. On the other hand, because the screening is imperfect, there is, as we saw in Chapter 4, at least in the Thomas–Fermi approximation, an exponentially damped Coulomb potential:

$$V(r) = e^2 \exp(-r/r_0)/4\pi\epsilon_0 r \tag{12.6}$$

Here r_0 is a screening radius of the order of the average interelectron separation, which in metals is of the order of the interionic spacing. This is called static screening because it depends only on distance and not on time. This interaction occurs in ordered or disordered systems but when, in the latter, the electrons are far apart diffusion alters the screening, which then depends on time. We call this dynamic screening and write the effective potential between electrons as $V(r, t)$ or, since it is usually more convenient to work with wavenumber q and frequency ω, in terms of the Fourier transform of the potential $V(q, \omega)$.

For simplicity we shall at first deal only with static screening even though we recognise that ultimately a correction will be needed. In section 12.5.1 we consider how the screening is changed by diffusion.

The Fourier transform of the static potential, equation (12.6), is:

$$V(q) = \chi^2/N(E_0)(q^2 + \chi^2) \tag{12.7}$$

where $\chi^2 = (r_0)^{-2} = e^2 N(E_0)/\epsilon_0$ with $N(E_0)$ the density of states at the Fermi level (see Chapter 4). As we noted earlier we are concerned with small values of q, where $V(q)$ is insensitive to the value of q. For later use, we note that when $q \to 0$, equation (12.6) assumes its maximum value:

$$V(0) = 1/N(E_0) \tag{12.8}$$

In this respect it is the same as the pseudopotential, which we discussed in section 4.6, whose value as q tends to zero is $-1/N(0)$, independent of the ion whose charge is being screened. We may also note that the $q = 0$ Fourier component is just the volume average of the interaction potential. We can see this immediately if, in equation (4.2a) for the Fourier transform, we put $\mathbf{K} = 0$ (equivalent to $q = 0$).

The interaction strength in the limit $q = 0$ (equation (12.8)) is thus independent of the electronic charge and depends, through the density of states, only on the screening medium.

12.3.2 The Hartree and exchange terms

This section is a digression to review the conventional treatment of the interaction between two electrons in the simplest approximation, where the wavefunction of the two electrons is written as a product of two single-electron wavefunctions. Quantum theory requires that the wavefunction of the two electrons should not distinguish one from the other and this requirement leads to the presence of two types of term in the final wavefunction, the Hartree and the exchange terms, as we now see.

If we consider the two electrons, one in a state of energy E at $\mathbf{r}$ and one in a state of energy E' at $\mathbf{r}'$, and if we use product wavefunctions, we must distinguish between electron pairs with parallel spins and those with antiparallel spins. The total wavefunctions for both sets have to be antisymmetrical, that is, they must change sign when the coordinates of the particles are interchanged. The space part of the product wavefunction integrated over space is symmetric and so, since with antiparallel spin pairs the spin part of the wavefunction is antisymmetric (see Table 11.2), the total wavefunction for such pairs is also antisymmetric. With the parallel spin pairs, however, the spin component is symmetric and so the space part of the wavefunction must be made antisymmetric. Thus we write for the space part of the wavefunction of a parallel spin pair of electrons:

$$\psi(\uparrow\uparrow) = \psi_E(\mathbf{r})\psi_{E'}(\mathbf{r}') - \psi_{E'}(\mathbf{r})\psi_E(\mathbf{r}') \qquad (12.9)$$

This changes sign if you interchange $\mathbf{r}$ and $\mathbf{r}'$. Moreover, as $\mathbf{r}$ approaches $\mathbf{r}'$ this tends to zero (reflecting the Pauli exclusion principle) so that there is a region around each electron which tends to exclude electrons of the same spin direction.

The interaction energy to first order in perturbation theory is:

$$E_{int}(\uparrow\uparrow) = \frac{1}{2}\int\int [\psi^*(\uparrow\uparrow) V(\mathbf{r} - \mathbf{r}', t)\psi(\uparrow\uparrow)]\mathrm{d}^3\mathbf{r}\mathrm{d}^3\mathbf{r}' \qquad (12.10)$$

where $\mathrm{d}^3\mathbf{r}$ and $\mathrm{d}^3\mathbf{r}'$ are volume elements at $\mathbf{r}$ and $\mathbf{r}'$ respectively, $V(\mathbf{r} - \mathbf{r}', t)$ the interaction energy and the integration is throughout the volume of the material. The separation of the two electrons is $\mathbf{r} - \mathbf{r}'$ and the factor $\frac{1}{2}$ prevents double counting when the result is summed over all electrons. If now from equation (12.9) we put into equation (12.10) the explicit form of $\psi(\uparrow\uparrow)$, we find two types of term: the first, the so-called Hartree term, has the form:

$$\int\int \psi_E(\mathbf{r})\psi_E^*(\mathbf{r})V(\mathbf{r}-\mathbf{r}',t)\psi_{E'}(\mathbf{r}')\psi_{E'}^*(\mathbf{r}')d^3\mathbf{r}d^3\mathbf{r}' \qquad (12.11)$$

The second, called the exchange term, has the form:

$$-\int\int \psi_E(\mathbf{r}')\psi_{E'}^*(\mathbf{r}')V(\mathbf{r}-\mathbf{r}',t)\psi_E^*(\mathbf{r})\psi_{E'}(\mathbf{r})d^3\mathbf{r}d^3\mathbf{r}' \qquad (12.12)$$

The Hartree term can be interpreted in analogy with a classical interaction: the terms $\psi_E\psi_E^*$ represent essentially the charge density at $\mathbf{r}$ and $\psi_{E'}\psi_{E'}^*$ that at $\mathbf{r}'$ so that equation (12.11) represents the energy of interaction due to appropriate charges at $\mathbf{r}$ and $\mathbf{r}'$.

By contrast, the exchange terms, which arise from making the total wavefunction of the electron pairs antisymmetric, cannot be interpreted classically. Their name arises from the fact that the exchange term, equation (12.12), resembles the Hartree term, equation (12.11), except that $\mathbf{r}$ and $\mathbf{r}'$ are interchanged in two of the wavefunctions so that in these E is associated with $\mathbf{r}'$ and E' with $\mathbf{r}$. As already noted the main purpose of the exchange terms is to ensure that when $\mathbf{r}$ approaches $\mathbf{r}'$, i.e. the two electrons get close to each other, the total wavefunction for parallel spin pairs become small and so prevents two electrons of like spin from coinciding.

Having now seen the form of the two types of term, Hartree and exchange, in the interaction energy of pairs of electrons, we now treat each type in turn because their contributions to the interaction energy are of quite different nature with different physical behaviour associated with each.

12.4 The exchange contribution

We begin with the exchange terms in the so-called particle–hole channel; the particle–particle channel, involving a different type of interaction, is dealt with in section 12.8.

These exchange terms in the interaction make an important contribution to the density of states because they are strongly energy dependent. The electron pairs that contribute most are, as we anticipated, those that interact through the small q components of the Thomas–Fermi potential. They are also those whose frequency difference ω is small. As we shall see, their diffusive motion makes possible access to states in k-space that ballistic electrons (electrons in free motion) cannot reach.

12.4.1 Calculation of the exchange interaction energy[3]

Let us begin by evaluating the contribution to the interaction energy of a typical exchange term:

$$-\int\int \psi_E^*(\mathbf{r})\psi_E(\mathbf{r}')V(\mathbf{r}-\mathbf{r}')\psi_{E'}(\mathbf{r})\psi_{E'}^*(\mathbf{r}')\mathrm{d}^3r\mathrm{d}^3r' \qquad (12.13)$$

which is just a re-ordered version of equation (12.12). We want to find the energy of interaction of two electrons, one at E_F (let this be E') and the other at an energy ϵ above it (let this be E).

As we saw earlier, conventional theory fails for electron pairs for which the scattering vector q is small; in order to identify such pairs we are obliged to work in reciprocal space and so we replace the potential in equation (12.13) by its Fourier transform:

$$V(\mathbf{r}-\mathbf{r}') = (2\pi)^{-3}\int V(q)\exp[i\mathbf{q}\cdot(\mathbf{r}-\mathbf{r}')]\mathrm{d}^3q \qquad (12.14)$$

We now separate the exponential so that the terms in $\mathbf{r}$ become:

$$\int \psi_E^*(\mathbf{r})\exp(i\mathbf{q}\cdot\mathbf{r})\psi_{E'}(\mathbf{r})\mathrm{d}^3r \qquad (12.15)$$

and those in $\mathbf{r}'$:

$$\int \psi_E(\mathbf{r}')\exp(-i\mathbf{q}\cdot\mathbf{r}')\psi_{E'}^*(\mathbf{r}')\mathrm{d}^3r' \qquad (12.16)$$

Given that r and r' are dummy variables, equations (12.15) and (12.16) are just complex conjugates and so their product is just the square modulus, which can be written in a different and more compact notation as:

$$|\langle\psi_E|\exp(i\mathbf{q}\cdot\mathbf{r})|\psi_{E'}\rangle|^2 \qquad (12.17)$$

where the bra $\langle\psi_E|$ and ket $|\psi_{E'}\rangle$ vectors imply integration over volume elements at $\mathbf{r}$ (see, for example, Cottrell 1988, p. 177). Equation (12.13) can then be rewritten as:

$$-(2\pi)^{-3}\int V(q)|\langle\psi_E|\exp(i\mathbf{q}\cdot\mathbf{r})|\psi_{E'}\rangle|^2\mathrm{d}^3q \qquad (12.18)$$

For values of q tending to zero, however, equation (12.18) becomes:

$$-(2\pi)^{-3}\int V(0)|\langle\psi_E|\psi_{E'}\rangle|^2\mathrm{d}^3q \qquad (12.19)$$

where the integral is for values of q for which $ql < 1$; the definition of l will become clear later.

This is an enormous simplification because we have now reduced a typical exchange term (in the range of small q) from a two-electron function in equation (12.13) to the one-electron function of (12.19). The square modulus of the scalar product or matrix element $\langle \psi_E | \psi_{E'} \rangle$ in (12.19) measures the probability that an electron will change its energy by ϵ (from E to E') and it is also, from the argument just given and under the conditions of our approximation, the probability that two electrons whose energies differ by ϵ should interact through their exchange terms.

12.4.2 Evaluation of $|\langle \psi_E | \psi_{E'} \rangle|^2$

One's first reaction is that $|\langle \psi_E | \psi_{E'} \rangle|^2$ must be zero unless $E = E'$ since the electron undergoes only elastic scattering and cannot change energy. This is indeed true of ballistic electrons but is no longer true of electrons that move by diffusion.

Such transitions must, of course, be virtual transitions, whose duration δt is short enough not to violate conservation of energy, i.e. δt is limited by the uncertainty principle so that $\epsilon \delta t < h$.

To go further we ought to use the Fourier transform of the classical diffusion probability to evaluate the square of the matrix element but to do so at this point would obscure the essential physics. Instead, let us first assume that the diffusive motion makes possible transitions from E' to E and that, to satisfy the uncertainty principle, the average electron lifetime in the higher energy state is proportional to but less than h/ϵ.

To evaluate $|\langle \psi_E | \psi_{E'} \rangle|^2$ for a single electron, we note that when $E = E'$ the normalisation of the wavefunctions ensures that $|\langle \psi_E | \psi_{E'} \rangle|^2 = 1$; otherwise it is zero. Therefore the value of $|\langle \psi_E | \psi_{E'} \rangle|^2$ averaged over time is proportional to the time spent in the higher energy state, i.e. to h/ϵ. Finally, according to equation (12.19) we must integrate the square of the matrix element over all the q states for which $ql < 1$, where l is the mean free path associated with the energy change ϵ. Call this mean free path L_ϵ. L_ϵ is thus the distance the electron diffuses during the lifetime of the state, i.e. in the time $\delta t \simeq h/\epsilon$. It is given by $(L_\epsilon)^2 \simeq D\delta t \simeq Dh/\epsilon$ or $L_\epsilon \simeq (Dh/\epsilon)^{1/2}$. q must therefore lie between zero and its maximum value $1/L_\epsilon = (\epsilon/Dh)^{1/2}$.

The integral over q gives the volume $4\pi q^3/3$ in k-space with q having its maximum value, so that this volume becomes $4\pi(\epsilon/hD)^{3/2}/3$.

Thus in equation (12.19), $|\langle \psi_E | \psi_{E'} \rangle|^2$ is proportional to $1/\epsilon$ and the integral over q is proportional to $(\epsilon/hD)^{3/2}$. The interaction energy thus varies as:

$$-V(0)(1/\epsilon)(\epsilon/hD)^{3/2} \tag{12.20}$$

This, with a factor of $N(0)$ and a reversal of sign, gives us the change in the density of states:

$$\Delta N(\epsilon) \propto N(0)V(0)\epsilon^{1/2}/(hD)^{3/2} \tag{12.21}$$

Thus the density of states varies as $\epsilon^{1/2}$ near the Fermi level.

The critical features in deriving this result are that: (1) the diffusive motion makes possible the transition from E' to E with a probability that is assumed independent of q and (2) the diffusive motion makes accessible a three-dimensional volume of k-space which is strongly energy dependent. To emphasise this last point, it is worth noting that for ballistic electrons, of energy $E = \hbar^2 k^2/2m$, a small change in energy ϵ causes a change in k, δk, given by $\epsilon = \delta E = \hbar^2 k \delta k/m$. With $\delta k = q$, we see that now $q \propto \epsilon$. Moreover, for ballistic electrons δk must lie in the direction of **k** and so, for electrons on a small area of Fermi surface, δS, the volume made accessible is qδS, which is proportional to ϵ. In the final result this would eliminate all dependence on ϵ.

12.4.3 The full effect of diffusion on $|\langle \psi_E | \psi_{E'} \rangle|^2$

The derivation of the result in equation (12.20) is, however, oversimplified in that it ignores the q dependence of the transition probability, $|\langle \psi_E | \psi_{E'} \rangle|^2$, which, as we shall soon see, is required by classical diffusion theory. We shall therefore now follow this fuller treatment to see this in detail.

We start with the expression (12.19) and assume that the probability for an electron in diffusive motion to change its wavenumber by q and its energy by $\epsilon = \hbar\omega$ is given by classical diffusion theory. This probability, denoted by $n(q, \omega)$, is the Fourier transform of the direct space probability:

$$P(r, t) = [\exp(-r^2/4\pi Dt)]/(4\pi Dt)^{3/2} \tag{12.22}$$

The required transform is:

$$n(q, \omega) = 1/(Dq^2 - i\omega) \tag{12.23}$$

or

$$n(q\omega) = (i\omega + Dq^2)/[\omega^2 + (Dq^2)^2] \tag{12.24}$$

This gives us the classical probability per unit frequency range of finding a particle with Fourier components q and ω; Figure 12.2 shows the shape

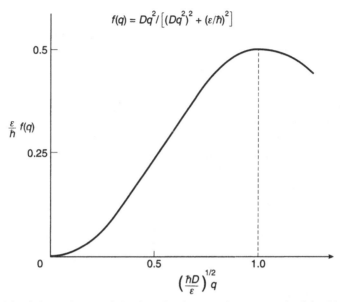

$$f(q) = Dq^2 / \left[(Dq^2)^2 + (\varepsilon/\hbar)^2 \right]$$

Fig. 12.2 If the real part of the function in equation (12.24) is $f(q)$, this shows how $(\epsilon/\hbar)f(q)$ varies with $(\hbar D/\epsilon)^{\frac{1}{2}} q$. Our oversimplified model corresponds to taking $(\epsilon/\hbar)f(q)$ as constant independent of q.

of the real part of this function in dimensionless form and the result (12.23) is derived in Appendix A3.

In classical terms this tells us nothing very surprising since it shows that, after a long time (small ω), the probability is high (large n) that a diffusing particle will be found at large distances (small q); the imaginary part indicates that the space and time parts are out of phase.

The physical meaning of equation (12.24) in our terms is that the diffusive motion of an electron alters its momentum states (k-states) in a rapid and irregular manner and so links them to the frequency response of the electron. The fact that the probability is complex implies that the transitions are virtual processes: their probability is given by the real part of equation (12.24) and the lifetime of the state by the imaginary part.

Our earlier oversimplified discussion is equivalent to assuming that the curve in Figure 12.2 is a constant independent of q instead of varying as shown: it starts from zero when $q = 0$ and reaches the value $1/2\omega = \hbar/2\epsilon$ when q reaches its upper limit $(\epsilon/\hbar D)^{1/2}$. Thus the average value of the real part over this range of q, which measures the probability of the transition, is proportional to $1/\epsilon$ as we assumed. Nevertheless the variation with q has to be taken into account in the integration over q and that is the reason why we need this fuller treatment.

We take the real part of equation (12.24) as giving the probability intensity (not amplitude because we are using a classical result) of finding a change q in wave number and $\hbar\omega$ in energy; the real part is thus equal to the square modulus in equation (12.19).

To use equation (12.24) in equation (12.19), however, we must convert it from unit frequency range to the probability per state. We first convert to unit energy range by means of Planck's constant and thence by means of the density of states to the probability per state. Finally we get:

$$|\langle\psi_E|\psi_{E'}\rangle|^2 = [\pi\hbar N(E)]^{-1}[Dq^2/\{(Dq^2)^2 + \omega^2\}] \tag{12.25}$$

We now put equation (12.24) into equation (12.19) with $\omega = \epsilon/\hbar$. We also multiply by $-N(0)$ to convert the expression from the interaction energy to the relative change in the density of states and find:

$$\Delta N(\epsilon)/N(0) = (2\pi)^{-3}N(0)V(0)\int[\pi\hbar N(\epsilon)]^{-1}[Dq^2/\{(Dq^2)^2 + (\epsilon/\hbar)^2\}]4\pi q^2\mathrm{d}q \tag{12.26}$$

where the volume element d^3q has been replaced by $4\pi q^2\mathrm{d}q$ appropriate to three dimensions; in one or two dimensions it would, of course, be different.

The density of states $N(\epsilon)$ can be taken as $N(0)$ and so the two cancel. In Figure 12.3 the integrand is shown in dimensionless form expressed as a function of the dimensionless variable $(\hbar D/\epsilon)^{1/2}q$.

The limits on q are as before: we require that $qL_\epsilon \leq 1$ and so q goes from 0 to $(\epsilon/D\hbar)^{1/2}$. Since we now know how important this upper limit is, we will discuss it more fully after we have finished the derivation.

To evaluate the integral we change the variable to $y = (\hbar D/\epsilon)^{1/2}q$ so that the limits are now 0 to 1. The change in the density of states apart from a numerical factor becomes:

$$(\epsilon/\hbar^3 D^3)^{1/2}N(0)V(0)\int_0^1 [y^4/(1 + y^4)]\mathrm{d}y \tag{12.27}$$

The definite integral gives a positive constant and the upshot is:

$$\Delta N(\epsilon) \propto N(0)V(0)\epsilon^{1/2}/(\hbar D)^{3/2} \tag{12.28}$$

which is the same result as equation (12.21) but now derived consistently.

Equation (12.28) is valid for both positive and negative values of ϵ and shows that there is a singularity in the density of states at the Fermi level.

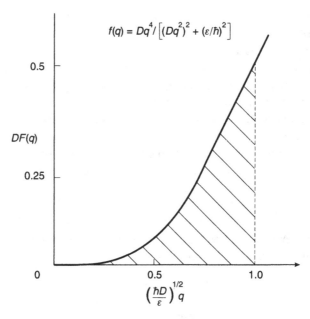

Fig. 12.3 Integrand of equation (12.26) as a function of q in dimensionless form.

The density of states varies as $|\epsilon|^{1/2}$ on each side of the Fermi level as shown in Figure 12.4.

As we shall discuss in more detail later, this means that, as the temperature rises from absolute zero, the electrons at the Fermi level, which are the only ones excited and which have an average thermal energy of about $k_B T$, sample more and more states at the higher density and so the density of states increases as $T^{1/2}$; the conductivity does likewise and causes the resistivity of the metal to fall with rising temperature.

Let us now return to the upper limit of q. It arises from the argument that, in interactions that involve large values of q, conventional theory is valid and these interactions make no contribution to the change in the density of states. The interactions that do change the density of states are those that involve small values of q where in conventional theory q would be ill defined. In this range of interaction we need to use a better approximation to the wavefunctions than plane waves and so, *faute de mieux*, we use classical diffusion theory to give us this better approximation. But this implies that we must stay in the realm of small ω and q, where classical theory should be valid. The upper limit of q thus marks the

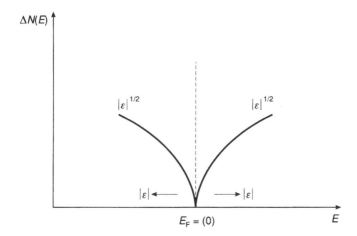

Fig. 12.4 Minimum in the density of states at the Fermi level due to the interaction effect (schematic).

boundary between the realm where our diffusion model is valid and that where conventional Boltzmann theory is appropriate. Moreover, it is not the precise magnitude of q at the boundary that is of prime importance but its dependence on ϵ. The appropriate mean free path that determines this limit on q is not the elastic mean free path: the critical distance for the transition between energy states is the mean distance the electron travels in the new state, which is the mean free path L_ϵ as derived above. The elastic mean free path l_0 is, of course, involved; it enters through the diffusion coefficient $D = v_F l_0/3$.

A final point is that the Hartree terms, of which equation (12.11) is an example, produce no similar contribution to the density of states because the time-dependent factors in the wavefunctions $\psi_E^*(\mathbf{r})$ and $\psi_E(\mathbf{r})$ are $\exp(-iEt/h)$ and $\exp(+iEt/h)$ respectively so that their product is independent of energy. Likewise for those in E'. Thus, although the energy of the electron states might be altered, their energy dependence would not. That is the reason for the importance of the exchange terms.

12.4.4 Summary

To find the density of states at an energy ϵ above the Fermi level we need to know how an electron at ϵ interacts with all the other electrons in the Fermi sea. In fact we need to know only how this energy differs from that of an electron in the state next higher in energy to ϵ so that ultimately we need only the interaction between the electron at ϵ and one at the Fermi level. The exchange terms, which intimately mix the wavefunctions of the two electrons, are the important ones with strong energy dependence. This energy dependence arises from electron pairs whose k-vectors differ by small values of q and whose frequency difference $\omega = \epsilon/\hbar$ is also small. Interaction depends on transitions between these energy states and is only possible because of diffusion. This diffusive motion not only alters the dependence of q on ϵ but also, unlike ballistic motion, allows q to take on all possible directions in space. Finally the density of states varies as $|\epsilon|^{1/2}$.

To complete the derivation of the change in density of states, we still have to calculate the dimensionless coupling constant $N(0)V(0)$ in equation (12.28). This we shall do in the next section.

12.5 Coupling constant for the exchange terms

To complete the evaluation of the contribution to the change in density of states from the exchange terms that we have been discussing, we must now determine the strength of the interaction in the $\omega = 0$ limit. We call this $V(0,0)$ to remind us that although we take the limiting value of $\omega = 0$, the screening is, because of diffusion, time dependent and differs from its static value, as we now discuss.

12.5.1 Diffusive screening

A physical argument to derive a measure of this type of screening is as follows. We consider the Fourier components q and ω of the interaction to be screened, which are therefore also the Fourier components of the charge distribution that the screening electrons must generate. A wave with these components propagates a distance $1/q$ in a time $1/|\omega|$. If the screening is instantaneous, this is also the screening time over this distance. (Incidentally the assumption of instantaneous screening is implicit in static screening, which shows that such screening, far from being static, is really superfast.) The time taken to travel the distance $1/q$ by an electron with diffusion coefficient D is $1/Dq^2$ so the total time of propagation and screening in the presence of diffusion is then $1/|\omega| + 1/Dq^2$. Hence the ratio of the time taken when diffusion is important to that when it is absent is:

$$(1/Dq^2 + /|\omega|)/(1/|\omega|) = (Dq^2 + |\omega|)/Dq^2 \qquad (12.29)$$

We see at once that the ratio is greater than or equal to unity: thus, as we would expect, any diffusive effects increase the screening time. Moreover for the small values of ω that concern us, fast diffusion (large values of D) or close encounters of the primary electrons (large values of q) take the ratio towards unity and diffusion makes no difference.

12.5.2 Size of the change in the density of states

For our purposes, we need the value of equation (12.29) in the limit that both q and $\omega \to 0$. But q and ω are related because the Fourier components of the interaction between the primary electrons are related. As we saw in the preceding section, for an energy change ϵ, q^2 must be less than or equal to $\epsilon/\hbar D$. Thus here with $\omega = \epsilon/\hbar$ we must have $Dq^2 \leq \omega$. Since this is true for the pair of primary electrons, it must also be true of the screening charge. Moreover there is another constraint on the response of the screening charge: the component with the largest q provides the fastest screening (see equation (12.29)) and so prevails. Thus the two conditions, (1) that q be a maximum and (2) that Dq^2 be less than or equal to ω, mean that $Dq^2 = \omega$. Thus ω and Dq^2 tend to remain equal as they both tend to zero. The limit of equation (12.29) is therefore 2, i.e. the screening time is doubled.

We now wish to find how this alters the screening radius. Because we are dealing with diffusive motion of the screening charge, we expect that the screening time will be proportional to the square of the screening radius, $(r_0)^2$ in the notation of equation (12.7), according to which:

$$V(q) = e^2/\epsilon_0(q^2 + \chi^2) \tag{12.30}$$

where $\chi^2 = (r_0)^{-2} = e^2 N(E_0)/\epsilon_0$.

Doubling the screening time means that the new screening radius is $2^{1/2}r_0$ and implies a halving of χ^2. Finally therefore we see from equation (12.7) above that, when $q = 0$, the interaction potential is:

$$V(0,0) = 2e^2/\epsilon_0\chi^2 = 2/N(0) \tag{12.31}$$

which is twice its static value.

(In highly disordered systems the screening of the pseudopotential should be similarly reduced so that the value of the pseudopotential at $q = 0$ and $T = 0$ would be doubled, i.e. $-2/N(0)$.)

Finally if we use this value of $N(0)V(0)$ and do a full calculation to determine the numerical constants, the result is:

$$\Delta N(0) = (|\epsilon|)^{1/2}/\pi^2(2\hbar D)^{3/2} \qquad (12.32)$$

which shows the characteristic $\epsilon^{1/2}$ dependence already discussed.

This completes our discussion of the exchange contribution to the density of states at absolute zero. We turn now to the Hartree contribution.

12.6 The Hartree terms in the particle–hole channel

12.6.1 The interaction mechanism

In the Hartree terms, as we saw in equation (12.11), each electron is associated with a wavefunction that describes in a quasi-classical fashion its own charge distribution and so we can describe the interaction in quasi-classical terms. The mode of interaction is, however, rather indirect and so let me first give an outline of it before we go into detail. This outline is based on the treatment by Bergmann (1987) of the enhanced interaction effect[4].

Under the conditions of strong elastic scattering there is, as we have seen in weak localisation, a substantial probability that a conduction electron will traverse a closed path and return to its starting point still retaining phase coherence. When this happens, the wavefunction can interfere with itself and set up a charge distribution that reflects the phase difference between the original and the returning wave. In the words of Bergmann this charge pattern forms 'an electron hologram', that is, a record of the phase change suffered by the electron in its closed path. For example, the returning wave could be exactly out of phase and so tend to cancel the original wave. At the other extreme, the returning wave could be exactly in phase so that the resultant would have its maximum possible amplitude. Likewise for intermediate phases. The important point is that this phase change can be inherited by another electron wave that passes through the charge hologram.

We imagine now that an electron and a hole execute the same closed path in opposite senses. On returning to its starting point, the electron, as we saw, leaves a charge record of the phase change that it has suffered on its path and the hole traversing the same closed path suffers the same phase change but as it passes through the charge hologram in the reverse sense this phase change is precisely reversed. This therefore restores the phase of its wavefunction to its original value and in this way the inter-action increases the amplitude of this second particle.

We can reinterpret this behaviour in terms of two electrons instead of an electron and a hole and since this is more familiar we adopt this description. We then have two electrons executing the same closed path in the *same* sense, with the second electron traversing the hologram in the reverse direction to that of the first.

The probability of executing the closed paths is determined entirely by the randomness of diffusion: the Coulomb interaction comes into play only when the second electron encounters the charge hologram left by the first.

The effect of this complex scattering mechanism is twofold: it contributes to a change in the density of states and it enhances the amplitude of the scattered wave. In this way the scattering via the Hartree interaction tends to increase the mean free path of one of the electrons involved and so leads to an increase in the conductivity compared to that expected by Boltzmann theory. We shall deal with this change of mean free path in the next chapter, which deals with the conductivity. In the meantime, we concentrate on the change in density of states.

12.6.2 Hartree contribution to the density of states

To determine how the Hartree terms in the particle–hole channel alter the density of states we must, as we saw in section 12.2, find the interaction energy of two electrons, one of energy ϵ and one at the Fermi level. The calculation thus involves two steps. First we must calculate the probability of forming suitable closed paths; second, we must calculate the inter-action energy involved when the two electrons interact at the site of the charge hologram. We take these in turn.

12.6.3 Probability of a closed path

We ask what is the probability of an electron executing a closed path through a succession of elastic scattering events and a second electron following this path. We answer this by treating the electrons as diffusing from site to site. Suppose that the probability amplitude of the first electron executing a closed path is A_1 and that of the second A_2. The combined probability intensity is $|A_1 + A_2|^2 = |A_1|^2 + |A_2|^2 + 2\text{Re}A_1A_2$. The first term describes the probability of the first electron describing a closed path, the second term that for the second electron. The cross terms refer to the combined probabilities that if electron 1 executes a closed path, electron 2 does likewise; they are the interference terms that depend on the phase of the two components and would not exist in a classical calculation. If the two amplitudes are in phase, the interference terms are both equal to A^2, which can be taken as the classical probability of a particle diffusing round a closed path, a quantity with which we are already familiar.

The second electron need not enter the closed path of the first electron at the same point; all that is required is that it should execute the same complete closed path and that, at the place where the hologram is, the second electron should traverse it in the appropriate sense. Any electron on the path of the first is a candidate for the interaction. Why does this imply that the second electron follows the same path as the first? The reason is that of all possible closed paths that start on the original path, only those at a phase extremum (which are just those we have chosen) survive; the others cancel each other out.

We now concentrate on the phase relationship of two electrons that must retain phase coherence throughout their closed paths. In addition to all the processes that can destroy phase coherence already discussed for weak localisation, we now have another effect. If the two electrons have different frequencies they cannot remain indefinitely in phase. Suppose that the two electrons in which we are interested differ in energy by ϵ, which corresponds to a frequency difference $\Delta\omega = \epsilon/\hbar$. If two waves start off in phase, they remain effectively in phase only for a time that is short compared to $1/\Delta\omega$; $1/\Delta\omega$ is roughly the time it takes to get completely out of phase due to the frequency difference. There will also be a difference in wave number Δk of the two electrons, which will alter their relative phase; this too is proportional to ϵ. From these effects we can define a time τ_ϵ, which is the time that elapses from when the two waves start out in phase until they get out of phase through their difference in energy. This gives $\tau_\epsilon \simeq h/\epsilon$, which is the time for A_1 and A_2 to get out of phase.

We must therefore determine the classical probability of completing a closed path in the time τ_ϵ. To do this, we turn to equation (11.5) for the classical diffusion time for a closed path and integrate it from τ_0, the elastic scattering time, to τ_ϵ, the available coherence time. This gives for the probability that our two electrons execute closed paths in phase:

$$p(\epsilon) \simeq \lambda^2 v_F[(\tau_0)^{-1/2} - (\tau_\epsilon)^{-1/2}]/(4\pi D)^{3/2} \qquad (12.33)$$

This shows that the probability of achieving suitable closed paths diminishes as $(\tau_\epsilon)^{-1/2}$ (i.e. as $\epsilon^{1/2}$) and the Hartree interaction between an electron at an energy ϵ above the Fermi level and one at the Fermi level does likewise. We must bear in mind, however, that this contribution only exists provided that there are no dephasing processes which on average operate in a time that is shorter than τ_ϵ.

How do we justify the use of classical diffusion in this case? In so far as the two waves are in phase the product of the two probability amplitudes gives a classical probability intensity for a closed path, which can therefore be calculated from the classical diffusion equation. Where the phases do not coincide, we take account of this failure by limiting the time over which the classical probability is calculated.

This brief description does not do full justice to the detailed derivation of the results (see Bergmann 1987) but it does give us enough physical insight to understand the factors that can destroy the interference, such as temperature, spin–orbit scattering and magnetic field, which we shall later discuss in some detail.

In the meantime, we turn to the second element in the interaction, which is the strength of the coupling between the two electrons that occurs in a localised region of space around the charge hologram.

12.6.4 The magnitude of the Hartree coupling strength

We must now work out the average energy of interaction of two electrons, both essentially at the Fermi level, which approach each other from random directions. We therefore evaluate this interaction by calculating its value when an electron is scattered from a given point on the Fermi surface to all other points and take the average. The Fourier transform $V(q)$ of the screened Coulomb potential can be written as (equation (12.7)):

$$N(0)V(q) = \chi^2/(q^2 + \chi^2) \tag{12.34}$$

where $N(0)$ is the density of states at the Fermi level and χ, the inverse of the screening radius, is given by:

$$\chi^2 = e^2 N(0)/\epsilon_0 \tag{12.35}$$

Our task is to find the average value of $V(q)$, given that the k-vector of the final state points into an element of solid angle $\sin\theta d\theta d\phi$ around the arbitrary direction θ, ϕ and the initial k-vector points in the z-direction. If the average of V over the complete solid angle 4π is written as $\langle V \rangle$ we have from equation (12.34):

$$N(0)\langle V\rangle = \int_0^{2\pi} \int_{-\pi/2}^{+\pi/2} \{4\pi[1 + (q/\chi)^2]\}^{-1} \sin\theta d\theta d\phi \qquad (12.36)$$

The integration over ϕ gives 2π and we then put $q = 2k_F \sin(\theta/2)$ with k_F the radius of the Fermi sphere. To integrate we write $\sin\theta d\theta = (1/2)d[\sin^2(\theta/2)]$. This gives:

$$2N(0)\langle V\rangle = [\ln(1 + x^2)]/x^2 \qquad (12.37)$$

where $x = 2k_F/\chi$. This interaction energy is conventionally denoted by F so that:

$$N(0)\langle V\rangle = F/2 \qquad (12.38)$$

If x is small, the function F tends to unity; if x is large, F becomes small. Therefore, if $\chi \ll 2k_F$, or, in words, if the screening radius is much bigger than the Fermi wavelength, F is small. This at first seems paradoxical because one would expect that if the screening radius is large, thereby increasing the effective range of the Coulomb interaction, the interaction energy would also be large. The reason why this is not so is that the Fourier components of the potential that come into play are those for which $0 < q < 2k_F$. If therefore q is limited to small values, equation (12.34) shows that only the large-amplitude components are involved whereas if q extends well above χ the small-amplitude components come to dominate the strength of the potential. Physically one can say that, if the screened potential varies rather slowly over several wavelengths of the electron wavefunction, the positive and negative regions can largely cancel. If on the other hand the strong potential is concentrated in a small part of the wave, the full effect shows itself.

12.6.5 The total contribution from the Hartree terms

We can now put together the two parts of our calculation to find the Hartree contribution to the density of states at absolute zero. This requires the product of the probability of suitable closed paths as given by the energy-dependent part of equation (12.33) with the coupling strength given in equations (12.37) and (12.38). Together these give:

$$\Delta N(\epsilon)/N(0) \simeq -F\lambda^2 v_F(\epsilon/\hbar)^{1/2}/(4\pi D)^{3/2} \qquad (12.39)$$

If now we write for the unperturbed density of states $N(0) = 3mk^2/(\pi h)^2$ and $\lambda = 2\pi/k_F$, we reach the final result:

$$\Delta N(0) \simeq -F\epsilon^{1/2}/(\hbar D)^{3/2} \qquad (12.40)$$

This holds for negative as well as positive values of ϵ. Thus the Hartree contribution to the density of states produces a singularity at the Fermi level varying as $|\epsilon|^{1/2}$. It is similar to that of the exchange contribution but of opposite sign (compare equations (12.11) and (12.12)) and with the additional factor F.

The Hartree contribution, and the exchange contribution given in equation (12.32), originate in quite different ways. The Hartree part comes from electron pairs executing closed paths and involves interference and coherence. In the exchange contribution the two electrons do not require coherence, only diffusive motion, similar energies and large separations.

The Hartree interaction is so subtle and intricate that one wonders if there may be other and perhaps even more elaborate modes of interaction that have not yet been identified. One can only rely on the systematics of theoretical methods and detailed comparison with experiment to reduce this possibility.

12.6.6 The importance of diffusion

We have now seen how the enhanced electron–electron interaction changes the density of states in a disordered metal in the neighbourhood of the Fermi level. Why is the singularity at the Fermi level?

Because we are dealing with a highly degenerate Fermi gas, the Fermi level is the natural reference level that separates the domain of electrons from that of holes. Alternatively we can say that it is only at this energy that there are enough occupied electron states with empty states nearby in energy to allow the transitions that the interaction induces (see the discussion of interactions and quasi-particles in Chapter 4). By itself, however, this factor is not enough to explain the change in density of states because highly degenerate ballistic electrons, interacting through a screened Coulomb potential, do not produce a singularity at the Fermi level. The other essential feature is the *diffusive* motion of the electrons. In diffusion there is a non-linear relation between the distance a particle travels and the time that elapses. The zero from which time is measured is thus critical. For ballistic motion the distance travelled depends only on the time *interval* and the zero of time does not matter. Thus only energy differences are important as we assumed at the outset for the unperturbed electron states (see Note 2).

When we convert this classical description of diffusion into quantum terms it means that, in the relation between energy and change of momentum **q**, the zero of energy (which is at the Fermi level) is now of critical importance. That is why there *is* a singularity at all and why it is found at that energy.

12.7 The combined contribution of Hartree and exchange terms

In evaluating the total coupling strength, we have to take proper account of the possible spin states of the two electrons. When this is done, the total change in the density of states due to the electron–electron interaction is:

$$\Delta N(0) = [|\epsilon|^{1/2}/\{2\pi^2(2hD)^{3/2}\}][2 - 3F/2] \qquad (12.41)$$

The term 2 in the second square bracket is the exchange term and the $3F/2$ the Hartree contribution; usually the exchange term dominates and if V is positive, as assumed in the simplest approximations, the combined effect is to produce a minimum in the density of states as illustrated in Figure 12.4. Equation (12.41) assumes that F is small. It also implies that we consider only the Coulomb repulsion between electrons and not the phonon-mediated attraction. To correct this we must replace $F/2$ by $(F/2) - \lambda_{ep}$ where λ_{ep} is the electron–phonon interaction parameter already introduced at the end of Chapter 7. Even then the expression for F in equation (12.37) has to be treated with some caution because the matrix elements of the interaction potential at short distances may require a better approximation to the real wavefunctions than plane waves.

When F is not very small compared to unity, we have then to take account of the fact that diffusion alters the screening of the interaction for small values of q (large separation of the electrons). This is already taken into account in the exchange term but not in the Hartree term, which consequently has to be modified to take account of the diffusion rate. It turns out that the form of the density of states is unchanged; only the coupling strength is slightly altered. Moreover this new coupling parameter can still be expressed in terms of F. The physical origin of this correction is clear and the final results are not greatly altered. They are given in Appendix A4.

12.8 Interactions in the particle–particle or Cooper channel

As we have frequently noted the interaction between electrons is not simply their mutual Coulomb repulsion, which has hitherto in this chapter been our main concern. There is also the phonon-mediated attraction that can lead to superconductivity. This is important for electron pairs with opposite k-vectors, which, as it were, interact head on.

The particle–particle channel, as distinct from the particle–hole channel, is concerned with just such electron pairs. As its name implies, it involves interaction between two electrons in states above the Fermi level but which, in addition, have equal and opposite momenta, that is, they are just the electron pairs that participate in superconductivity at low enough temperatures. The important point, however, is that this interaction is significant well above the superconducting transition temperature and, even if it is not strong enough to cause superconductivity at all, can still influence the density of states in a disordered system. The electron pairs are not unlike Cooper pairs, although they do not have to have opposite spin directions, and one may perhaps think of their effect as that of enhanced, fluctuation-induced Cooper pairs. For obvious reasons this channel is also referred to as the Cooper channel.

To calculate the change in the density of states, we proceed in the same way as for the particle–hole channel. As before, the calculation involves finding the energy of interaction of an electron at the energy of interest (at an energy ϵ above the Fermi level) and an electron at the Fermi level. The difference between the Cooper channel and the Hartree terms in the particle–hole channel is that now the two electrons execute closed paths in opposite senses instead of in the same sense. As before the interaction energy depends on the product of two factors: the classical probability of executing a closed path and the interaction energy between the two oppositely directed electrons. The calculation of the probability of closed paths is the same as in our previous calculations and gives rise to the same $\epsilon^{1/2}$ dependence for the density of states. Now, however, the interaction at the charge hologram has quite a different form. It does not involve the parameter F but rather the electron–phonon coupling parameter λ_{ep}, which we have already discussed.

We therefore concentrate only on this aspect of the Cooper channel. The interaction is essentially that which produces superconductivity and so it is not surprising that it involves a competition between the energy of the electron ϵ and an energy characteristic of superconductivity. For an electron with energy ϵ above the Fermi level, the coupling strength is of the form:

$$1/\ln(k_B T_c/\epsilon) \tag{12.42}$$

where, if the material is a superconductor, T_c is the superconducting transition temperature. The rather strange form of equation (12.42) is perhaps explained if one recalls from the theory of superconductivity that T_c can be written:

$$T_c = \theta_D \exp(1/\lambda_0) \quad \text{or} \quad \lambda_0 = 1/\ln(T_c/\theta_D) \qquad (12.43)$$

where θ_D is the Debye temperature and λ_0 is the bare electron–phonon interaction parameter (here λ_0 is negative because the material is a superconductor). Thus if we put $\epsilon = k_B\theta_D$ in equation (12.42), the coupling strength is just λ_0.

If the net interaction between electrons is repulsive (λ positive) so that there is no superconductivity, equation (12.43) takes the form:

$$1/\ln(k_B T_F/\epsilon) \qquad (12.44)$$

where T_F is the Fermi temperature of the electrons. These two definitions of T_c reflect the fact that for the superconducting material the range of interaction is limited by the phonon energies whereas the Coulomb repulsion is limited only by the Fermi energy.

Given that the interaction in the Cooper channel involves electron pairs of opposite momenta for which $\mathbf{k} + \mathbf{k'} \simeq 0$, there is no exchange contribution comparable to that in the particle-hole channel, which required that $\mathbf{k} - \mathbf{k'} \simeq 0$. In the Cooper channel the electrons tend to interact at close range so that the exchange terms simply tend to offset the Hartree terms with parallel spin (see equations (12.9) and (12.12) and the discussion there). Indeed if the bare electron–phonon interaction λ_0 does not depend on energy or momentum transfer (ω and q), then the exchange terms will almost completely cancel these parallel spin Hartree terms; if the interaction is a δ-function, the cancellation is complete. Thus the only remaining contribution is from Hartree terms of opposite spin pairs.

The total change in the density of states at absolute zero, when F is not very small and when the correction for diffusive screening is included, then becomes:

$$\Delta N(0) = [|\epsilon|^{1/2}/\{\pi^2(2hD)^{3/2}\}][1 - 3\{(1 + F/2)^{1/2} - 1\} \\ - 1/\ln(k_B T_c/|\epsilon|)] \qquad (12.45)$$

If the value of F is small and we neglect the log term, the coupling strength reduces to that in equation (12.41).

12.9 Change of density of states with temperature

When the temperature T is raised above absolute zero, the change in the density of states due to the interaction effect depends on both temperature and ϵ. If we consider states at an energy $\epsilon \ll k_B T$, we may expect that they will be perturbed by temperature because their self-energy depends on electron pairs with frequencies that differ by only $\epsilon/\hbar$. Their interaction will be modified in two ways: first, some of the virtual processes can become real through thermal excitation and, second, some virtual processes will be inhibited because the states into which the virtual process would take the electron are now occupied.

The density of states in this region of energy thus tends to lose its dependence on ϵ but become dependent on the temperature. In particular this is true of the density of states around the Fermi level, which is of special concern to us since it is this that determines, at least in part, how the conductivity itself varies with temperature.

If, as is usual, the total coupling strength is dominated by the exchange terms in the particle–hole channel and so is positive, there is, as we saw, a minimum in the density of states at the Fermi energy. The main effect of temperature is to average the density of states over the range of energies $k_B T$ around E_F and, because of the minimum in the density of states at E_F, the average value of $N(E_F)$ will rise as the temperature rises and as the electron distribution samples more and more states with higher density. Given that $\Delta N(\epsilon)$ at absolute zero varies as $|\epsilon|^{1/2}$ in this region, we can expect that the average of $\Delta N(E_F)$ will vary as $T^{1/2}$. Thus we have:

$$N(E_F) = A + B(k_B T)^{1/2} \qquad (12.46)$$

where A and B are positive constants.

Finally let us note that the change in density of states has a direct effect on the conductivity as is clear from the Einstein relation, which in differential form tells us that:

$$\Delta\sigma/\sigma = \Delta N(E_F)/N(E_F) + \Delta D/D \qquad (12.47)$$

where D is the diffusion coefficient. (We assume the validity of this expression even in the presence of interactions.)

As a first approximation we can say that weak localisation alters the diffusion rate (the second term on the right) but leaves the density of states unchanged whereas the enhanced electron–electron interaction alters the density of states but leaves the diffusion coefficient almost untouched. In fact, as we have hinted earlier, the interaction does alter the scattering and hence D, as we discuss in the next chapter.

12.10 Summary

The interaction of electrons diffusing through a disordered metal causes a singularity in the density of states at the Fermi level which in turn alters the temperature dependence of the conductivity.

There are three main contributions. The most important is from the exchange terms associated with electron–hole pairs that interact with small momentum change q. In the realm where $q \Rightarrow 0$, the screened Coulomb interaction is at its maximum. Moreover the diffusive motion of the electrons makes possible and amplifies interactions between electrons with different energies, in particular by making accessible regions of k-space that are inaccessible to ballistic electrons.

The other two contributions involve electron pairs that execute closed paths; their wavefunctions interfere in such a way that the interaction depends on energy and so alters the density of states. In one case the electrons interact through the screened Coulomb force and in the other, the so-called Cooper channel, through the phonon-mediated interaction that can lead to superconductivity. All three contributions cause the density of states to vary as the square root of the energy difference from the Fermi level.

13

The effect of the Coulomb interaction on conductivity

13.1 Introduction

We have now seen how the enhanced interaction effect changes the density of states in the neighbourhood of the Fermi level. In this chapter we consider how this interaction alters the conductivity, first by contributions from the exchange and Hartree terms in the particle–hole channel and then those from the Cooper channel.

13.2 The particle–hole channel

13.2.1 Exchange terms

The exchange part of the Coulomb interaction in the particle–hole channel makes its contribution to the change in conductivity essentially through the change in density of states. Let us estimate the size of the effect from these terms at a temperature of, say, 1 K in a metallic glass. We use equation (12.32) and put $\epsilon = k_B T$ with $T = 1$. We take $D = 5 \times 10^{-5}\,\mathrm{m^2\,s^{-1}}$. For the density of states at the Fermi level in, say, 50–50 CuTi, we take $2\frac{1}{2}$ conduction electrons per atom and a Fermi energy about 5 eV above the bottom of the band. The density of states is thus about 0.5 states per eV per atom, which, converted to MKS units, gives about 10^{47} states $\mathrm{J^{-1}\,m^{-3}}$. The final estimate of the relative change in the density of states, and hence of the conductivity, is of the order of 1 %.

When properly calculated, the exchange contribution to the change in conductivity is:

$$\Delta\sigma = 0.915[e^2/3\pi^2\hbar](k_B T/\hbar D)^{1/2} \tag{13.1}$$

which in relative terms corresponds quite closely to the change in the density of states that arises from the exchange terms in the particle–hole channel.

13.2.2 Hartree terms

The mechanisms that underlie the Hartree terms in the interaction have already been described but let me summarise them briefly. An electron, under the influence of elastic scattering from the disordered ions, diffuses round a closed path and on returning to its starting point leaves a charge distribution, which, like a hologram, records the change in phase of the wave in going round the circuit. A hole traverses the same circuit in the opposite sense, thereby suffering the same phase change as the first. It can then traverse the charge hologram in the opposite sense to the electron and so recover its initial phase. The effect of this complex scattering mechanism is thus, as we saw, to amplify the amplitude of the scattered wave so that it tends to increase the mean free path of the hole involved and so increase the conductivity. The effect is thus opposite to that of the change in density of states.

Instead of a hole executing the closed path in the opposite sense to the electron it is easier to think of the equivalent electron traversing the closed path in the same sense as the first electron and having its phase restored by the hologram by time reversal.

To estimate the change in conductivity brought about by these scattering processes, we need to know how the amplitude of the wavefunction of the second electron is altered. An outline argument is as follows. The first electron starts with unit amplitude and after completing its closed path has amplitude A_1. This determines the intensity of the charge hologram. The second electron likewise sets out and completes the circuit, ending with amplitude A_2 and now interacts with the hologram charge with a scattering amplitude $\mathscr{S}_2$. Thus the final amplitude is $1 \times A_1 \times A_2 \times \mathscr{S}_2$ since all the processes are in phase. The two amplitudes A_1 and A_2 both refer to the same path and are in fact the same, so that their product is just $|A|^2$ and is the probability of completing the closed path; this is taken from the classical diffusion probability.

$\mathscr{S}_2$ is the scattering amplitude for an electron scattered by the screened Coulomb potential of another electron; in the Born approximation, if q is the magnitude of the scattering vector, the scattering amplitude is just the q-component of the Fourier transform of the scattering potential. This is given in equation (12.34). If we average this for values of q that describe all scattering processes on the Fermi surface, we are in fact repeating the calculation already carried out in section 12.6.4; we thus get the quantity F as evaluated there.

We have assumed that the second electron completes its closed path before interacting with the other electron but this is not necessary; it can enter the closed path at any point, interact at the place of the hologram and then complete its circuit.

Thus the change in the scattering probability of the second electron is given by the product of F and the diffusion probability for a closed path. Before we can estimate this, however, we must know how long the two electrons maintain phase coherence. In general this is destroyed by the effects of temperature as we now see.

When we considered the Hartree contribution to the density of states, we found that an energy difference between the two electrons produces a phase difference that increases with time. At a temperature T this energy difference can be of order $k_B T$. Moreover, the nature of the interaction we have just discussed implies that the electrons involved are well localised and this in turn means that their wavefunctions must correspond to quite well-defined wave packets. Such packets are composed of waves from a range of frequencies centred on the one that dominates and the range of frequencies involved depends on the thermal spread of electrons around the Fermi level, i.e. also on $k_B T$. Consequently a typical difference in energy between waves is of this order and their *frequency* difference of order $\Delta\omega = k_B T/\hbar$. Two waves with a frequency difference $\Delta\omega$ get completely out of phase in a time of order:

$$\tau_T \simeq 1/\Delta\omega = \hbar/k_B T \qquad (13.2)$$

which is the thermal coherence time. At $10\,\text{K}$ this has a value of about 10^{-13} s, which is long compared to a typical elastic scattering time in a highly resistive alloy. (Incidentally, the reason why thermal incoherence does not influence the mechanism of weak localisation is that in this case each harmonic component of the electron wavefunction interferes with itself.)

To determine the required probability of the two electrons executing a closed path, we turn to equation (11.5):

$$p(t)dt \simeq \lambda^2 v_F dt/(4\pi Dt)^{3/2} \qquad (13.3)$$

Now we assume that the diffusion processes are terminated by thermal incoherence rather than by inelastic scattering and so we integrate from τ_0, the elastic scattering time, to τ_T, the thermal coherence time, to give us the appropriate probability. Moreover to get the relative change in mean free path and hence conductivity, we multiply by F to get:

$$\Delta\sigma/\sigma \simeq F\lambda^2 v_F[(\tau_0)^{-1/2} - (\tau_T)^{-1/2}]/(4\pi D)^{3/2} \qquad (13.4)$$

The first term in the square brackets is constant but the second is temperature dependent and shows that the probability of achieving suitable closed paths diminishes as $(\tau_T)^{-1/2}$ i.e. as $T^{1/2}$. This means that the probability of enhancing the electron wavefunction through the Hartree interaction diminishes as $T^{1/2}$ at low temperatures and that the corresponding part of the conductivity decreases with the same power law. If now we concentrate on the temperature-dependent part of equation (13.4) and we put $\sigma = (ek_F)^2 l/3\pi^2 \hbar$ with $\lambda = 2\pi/k_F$, we find:

$$\Delta\sigma \simeq -e^2 F(k_B T/\hbar D)^{1/2}/6\pi^{3/2}\hbar \qquad (13.5)$$

This is the required result except that the numerical constant is not reliable and we must bear in mind that the Hartree contribution only exists provided that there are no additional dephasing processes to destroy the phase coherence of the two electrons.

Let us write equation (13.5) as:

$$\Delta\sigma = -\text{constant } F(e^2/h)(k_B T/\hbar D)^{1/2} \qquad (13.6)$$

which refers to the interaction effect, and compare it with equation (11.10):

$$\Delta\sigma = \text{constant } (e^2/h)(1/\tau_{in}\hbar D)^{1/2} \qquad (13.7)$$

which refers to weak localisation. As we have seen, this very close resemblance is not accidental.

13.2.3 Total change in conductivity from the particle–hole channel

A full calculation of the contributions to the change in conductivity from both exchange and Hartree terms in the particle–hole channel gives:

$$\Delta\sigma(T) = 0.915[e^2/4\pi^2\hbar]\{(4/3) - 3\tilde{F}/2\}(k_B T/\hbar D)^{1/2} \qquad (13.8)$$

where the first term in the curly brackets is from the exchange and the second from the Hartree processes; they have opposite signs because the density of states increases as T increases and so raises the conductivity while the Hartree enhancement of the conductivity is reduced by the rising temperature. Since in general the exchange terms tend to dominate, the conductivity rises as $T^{1/2}$ and so provides another mechanism for a negative temperature coefficient of resistance. Such a temperature dependence with this small power law ($T^{1/2}$) tends to prevail at the lowest temperatures and is indeed so found in the conductivity of metallic glasses; it is thus attributed to the interaction effect.

The coupling strength in the Hartree processes is not exactly F but $\tilde{F}$. The change in value, which is given in Appendix A4, comes about because the interaction between the electron pairs is altered by the diffusive motion of the screening electrons as was the contribution of these pairs to the density of states (section 12.5.1).

What we have seen so far is valid in the absence of spin–orbit scattering but the contribution involving $\tilde{F}$ in equation (13.8) is strongly influenced by such scattering, as we now discuss.

13.3 Enhanced interaction effect and spin–orbit scattering

The influence of spin–orbit scattering on the interaction effect does not appear to have been treated in the literature but we can gain a qualitative picture of its influence in the same way that we discussed its influence on weak localisation in Chapter 11. There the spin–orbit scattering destroyed the phase relationship between partial waves of *like* spin electrons executing closed paths in *opposite* senses: here by a similar argument, spin–orbit scattering destroys the phase relationship of electron pairs of *opposite* spin executing paths in the *same* sense. Moreover of the two contributions to the conductivity, $\Delta\sigma_{ex}$ and $\Delta\sigma_{H}$, that come from exchange and Hartree terms, only $\Delta\sigma_{H}$ is altered by spin–orbit scattering. $\Delta\sigma_{ex}$ is unaltered because it does not depend on interference.

The two contributions, of opposite sign, but, in the absence of spin–orbit scattering, both varying as $T^{1/2}$, are shown, measured from the Boltzmann value as origin, in Figure 13.1(a). Plotted against $T^{1/2}$ they are straight lines.

We can treat spin–orbit scattering as a random dephasing mechanism, analogous to temperature, except that it acts selectively only on the antiparallel spin pairs. Spin–orbit scattering thus reduces the value of $\Delta\sigma_{H}$ at absolute zero and keeps it constant until, with rising temperature, thermal incoherence begins to rival the rate of spin–orbit scattering and, having overtaken it, causes $\Delta\sigma_{H}$ to return towards the value it would have without spin–orbit scattering. These effects are shown for several increasing values of this scattering in Figure 13.1(a); in the limit, $\Delta\sigma_{H}$ is reduced to zero.

Figure 13.1(b) shows $\Delta\sigma_{total}$, which is the algebraic sum of the two parts; it shows that $\Delta\sigma_{total}$ starts at absolute zero with the slope of $\Delta\sigma_{ex}$,

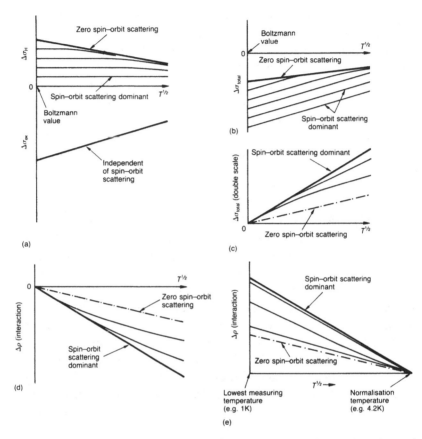

Fig. 13.1 The temperature dependence of the conductivity due to the enhanced interaction effect in the presence of spin–orbit scattering (schematic). (a) The Hartree contribution $\Delta\sigma_H$ and the exchange contribution $\Delta\sigma_{ex}$ to the conductivity with respect to the Boltzmann value as zero plotted against $T^{1/2}$. $\Delta\sigma_{ex}$ is unchanged by spin–orbit scattering but $\Delta\sigma_H$ is gradually reduced as the strength of this scattering is increased (successively lower curves in the diagram). (b) The total conductivity change, being the sum of $\Delta\sigma_{ex}$ and $\Delta\sigma_H$. Since in general $|\Delta\sigma_{ex}| > |\Delta\sigma_H|$, the slope of $\Delta\sigma_{total}(T)$ is always positive but the slope increases as the spin–orbit scattering increases. When this scattering dominates throughout the temperature range, only $\Delta\sigma_{ex}$ contributes to the total. (c) The total conductivity change measured from a common origin for which $\Delta\sigma(T)$ is zero when $T = 0$. (d) The same as in (c) but plotted as change in resistivity. (e) The same data as in (d) over a limited temperature range and normalised at some upper temperature (say, 4.2 K) as done with experimental data.

except in the physically unrealistic case of zero spin–orbit scattering. As the spin–orbit scattering increases, the curves approach closer and closer to the lowest line in Figure 13.1(b), which is $\Delta\sigma_{ex}$ alone and corresponds to a sample in which spin–orbit scattering is dominant throughout the temperature range plotted.

Finally Figure 13.1(c) shows all curves starting from a common origin at absolute zero. Notice that the slope, while always positive, is altered by the spin–orbit scattering and becomes smaller at temperatures where τ_{so} becomes comparable to τ_T. We shall see in section 16.4 how these predictions compare with experiment.

Figure 13.1(d) shows the corresponding changes in resistivity ($\Delta\sigma/\sigma = -\Delta\rho/\rho$) and Figure 13.1(e) shows what happens when the curves are normalised at some specific temperature (not absolute zero). This is done with experimental data because it is not possible to measure $\Delta\sigma$ in absolute terms.

We can make the previous argument quantitative in the following way. In equation (13.8), the second term in the curly bracket, involving $\tilde{F}$, is altered by spin–orbit scattering. The electron pairs that contribute to it are put out of phase not only by thermal incoherence in a time $\tau_T = \hbar/k_B T$ but also by the competing mechanism of spin–orbit scattering in a characteristic time τ_{so}. The combined probability of scattering is thus:

$$1/\tau_{total} = 1/\tau_T + \alpha/\tau_{so} = k_B T/\hbar + \alpha/\tau_{so} \qquad (13.9)$$

where α is a weighting factor, or order unity to take account of the degeneracy of the spin states and the definition of τ_{so}.

This means that the temperature involved in the coefficient of $\tilde{F}$ must be replaced by an effective temperature:

$$T_{eff} = T + T_{so} \qquad (13.10)$$

where $T_{so} = \alpha\hbar/k_B\tau_{so}$. If we put this in equation (13.8) we get:

$$\Delta\sigma(T) = 0.915(e^2/4\pi^2\hbar)[(4/3)(k_B T/\hbar D)^{1/2} - (3\tilde{F}/2)\{(k_B T_{eff}/\hbar D)^{1/2} - (kT_{so}/\hbar D)^{1/2}\}]$$

$$(13.11)$$

with T_{eff} given by equation (13.10). The second term in the curly brackets ensures that $\Delta\sigma$ is zero at $T = 0$.

In the limit that $\hbar/\tau_{so} \gg k_B T$, T_{eff} becomes effectively independent of T so that the curve of $\Delta\sigma$ versus $T^{1/2}$ is a straight line whose slope is determined by the factor $(4/3)(k_B T/\hbar D)^{1/2}$. In the other limit where $\hbar/\tau_{so} \ll k_B T$, the curve is again a straight line but now the slope is determined by the factor $[(4/3) - (3/2)\tilde{F}](k_B T/\hbar D)^{1/2}$. Thus the ratio of these two extreme slopes provides a way of finding $\tilde{F}$.

A further consequence of spin–orbit scattering is that, at temperatures where neither the spin–orbit scattering nor thermal incoherence dominates, the change in conductivity is no longer linearly proportional to $T^{1/2}$; the plots of $\Delta\rho$ versus $T^{1/2}$ have a positive curvature.

13.4 The Cooper channel: additional effects

As we have already discussed in Chapter 12, additional contributions from the Cooper channel arise from the electron–phonon interaction which promotes superconductivity. The interaction in the Cooper channel can be thought of as similar to that of the Hartree terms of the particle–hole channel except that now the two electrons travel round the same closed path in opposite senses and thus interact with each other, as it were, head on as in the superconductive interaction.

The superconducting fluctuations in the neighbourhood of the superconducting transition temperature T_c are well known and do not concern us here. As we have seen, however, the fluctuation-induced formation of Cooper pairs at temperatures far from T_c causes a change in the density of states and this influences the conductivity even in alloys that do not become superconducting.

The change in conductivity at temperatures much greater than T_c is:

$$\Delta\sigma_c = -0.915[e^2/2\pi^2\hbar](k_B T/\hbar D)^{1/2}[\ln(T_c/T)]^{-1} \tag{13.12}$$

The temperature dependence is dominated by the $T^{1/2}$ term and is thus similar to that in equation (13.5). The sign of the contribution depends on the sign of the interaction between the electron pair at short distances. It makes the conductivity grow with increasing temperature if attractive and vice versa. The relation between equation (13.12) and the last term of the density of states expression in equation (12.45) is clear.

Another manifestation of the superconducting fluctuations is the so-called Maki–Thompson correction, which is small compared to that in equation (13.12) but is particularly important in the presence of a magnetic field (Appendix A5).

13.5 Summary

The enhanced interaction effect alters the conductivity of highly disordered alloys in three main ways, which have similar temperature dependences.

1 The exchange terms in the particle–hole channel decrease the density of states and hence the conductivity at absolute zero and cause it to rise as $T^{1/2}$ as the temperature goes up.

2 The Hartree contribution in the particle–hole channel arises from electron–hole pairs which interfere in such a way as to increase their mean free path and hence the conductivity at absolute zero. Thermal incoherence destroys the interference and causes the conductivity to fall as $T^{1/2}$. The interference also alters the density of states in a sense opposite to that in 1 above.

3 Electrons in the Cooper channel alter the density of states through the phonon-mediated attraction between electrons, the sign depending on that of the net interaction between electrons. There is also an effect on the mean free path analogous to that in 2.

14

Influence of a magnetic field on the enhanced interaction effect

14.1 Magnetoresistance in the particle–hole channel

14.1.1 High fields

When a magnetic field is applied to the material, the relative phase of the two electrons in the processes we discussed in section 13.2.2 is not changed by the flux that passes through their common orbit because the electrons execute this in the same sense (not in opposite senses as in weak localisation) and so the effect on the phase is the same for both. On the other hand if the two electrons have antiparallel spins the magnetic field B changes their relative energy by the Zeeman splitting $g\mu B$. Here μ is the Bohr magneton and g is the splitting factor, which looks after any change in the magnetic moment of the electron introduced by its environment. In fact we are here considering electron–hole pairs so that since the hole has a spin opposite in sign to that of an electron the triplet state occurs when the electron and hole have antiparallel spins and the singlet state when they are parallel.

If we refer to Table 11.2 (p. 126), we see that in the triplet state only two out of the three spin wavefunctions involve parallel spins; the third is a composite state, which, like the singlet state, involves antiparallel spins. Thus the two components of the electron–hole pair with parallel spins are the ones that are split in energy by the magnetic field. The frequencies of the electron and the hole are correspondingly altered and for this reason dephasing occurs. Thus the contribution to the anomalous resistance from two of the triplet states is reduced. On the other hand the contributions from the other wavefunctions are unchanged by this mechanism, including those electrons described by the exchange terms, and being uninfluenced by the magnetic field they can be ignored in this discussion.

When the magnetic field causes a dephasing of the electron–hole pairs this diminishes the Hartree contribution. As we saw earlier, the Hartree term enhances the amplitude of the wavefunction and so increases the conductivity of the material. If therefore we apply a magnetic field and reduce this contribution we diminish the conductivity, i.e. we produce a positive magnetoresistance, which is thus opposite in sign to that of weak localisation. (Remember, however, that spin–orbit scattering can reverse the sign of the weak localisation effect.)

The field dependence of the magnetoresistance at low temperatures due to the interaction effect can be derived from the above argument. The energy difference $g\mu B$ induced by the magnetic field produces a difference in frequency of $g\mu B/\hbar$ between the two relevant components. They thus get out of phase in a time τ_B of order $\hbar/g\mu B$. Of course there are closed paths of different lengths with different completion times so that there is a spectrum of phase differences among the different pairs that complete closed paths, some greater and some smaller than that corresponding to τ_B. Thus some pairs will tend to increase and some to decrease the resistivity and we take τ_B to be the time taken for the total contribution from these terms to be reduced to some fraction of its initial value.

To find out how this alters the resistance we proceed as we did for thermal incoherence; here we integrate equation (13.3) from τ_0 to τ_B to determine the probability of the electrons (we think now in terms of two electrons instead of the electron and hole) returning to their starting point before the magnetic field causes them to cease contributing. The analogue of equation (13.4) then becomes:

$$p(B) \simeq \lambda^2 v_F [(\tau_0)^{-1/2} - (\tau_B)^{-1/2}]/(4\pi D)^{3/2} \qquad (14.1)$$

The relative change in conductivity is found by multiplying by F:

$$\Delta\sigma/\sigma \simeq -[F\lambda^2 v_F/(4\pi D)^{3/2}][(\tau_0)^{1/2} - (\mu g B)^{1/2}] \qquad (14.2)$$

and hence, by putting $\sigma = (e^2/3\pi^2\hbar)k_F^2 l$, $\lambda = 2\pi/k_F$ and $D = v_F l/3$, we get for the change in conductivity $\Delta\sigma$ due to the appropriate Hartree terms, apart from a numerical constant:

$$\Delta\sigma(B) \simeq -F(e^2/2\pi^2\hbar)(g\mu B/\hbar D)^{1/2} \qquad (14.3)$$

This is valid at absolute zero; the behaviour at higher temperatures is discussed below.

14.1.2 Low fields

What we have discussed so far is strictly valid at absolute zero but is approximately true at fields for which $g\mu B \gg k_\mathrm{B}T$. At small fields where $g\mu B \ll k_\mathrm{B}T$ the magnetoresistance has a different field dependence. Because the field is so small, the coherent lifetime of the electrons that interact is limited, as in zero field, by the temperature but in addition each electron that interacts with a suitable partner differs from it in energy by the Zeeman splitting $g\mu B$. The two electrons thus progressively get out of phase with a difference of $g\mu Bt/\hbar$ after time t. They lose coherence after the thermal coherence time $\tau_T = \hbar/k_\mathrm{B}T$ by which time the phase difference δ due to the magnetic field is $\delta = g\mu B/k_\mathrm{B}T$. This is true of *all* the pairs that respond to the field so that whereas the influence of temperature is random, that of the field is the same for all. Consequently, as we saw earlier in our discussion of the low-field magnetoresistance due to weak localisation, the intensity of the coherent contribution is reduced by a factor $\cos^2(\delta/2)$. The magnetoconductivity thus is proportional to:

$$[1 - (g\mu B/2k_\mathrm{B}T)^2] \tag{14.4}$$

These are the first two terms in the expansion of $\cos^2(\delta/2)$ and we see therefore that the magnetoresistance $\Delta\rho/\rho = -\Delta\sigma/\sigma$ is positive and varies in this low-field regime as $(g\mu B/k_\mathrm{B}T)^2$.

14.2 Temperature dependence of magnetoresistance

To demonstrate more clearly how the conductivity depends on the magnetic field and temperature, Figure 14.1(a) shows the Hartree contribution to the conductivity in relation to the Boltzmann value, which we take to be independent of temperature or magnetic field.

The figure shows $\Delta\sigma$ at different temperatures as a function of $B^{1/2}$. The uppermost curve is for absolute zero and, in the absence of spin–orbit scattering, which we treat below, is a straight line since, under these conditions, the conductivity falls as $B^{1/2}$ even down to the lowest fields. At higher temperatures, as shown in the figure, the value in zero field $\Delta\sigma(T,0)$ is reduced by thermal incoherence and, as the field is turned on, the conductivity is still further reduced, varying as B^2. Eventually when $\mu B \gg k_\mathrm{B}T$ all the curves asymptotically approach the absolute zero curve as shown because ultimately the magnetic field dephases the electron pairs before thermal incoherence has time to operate.

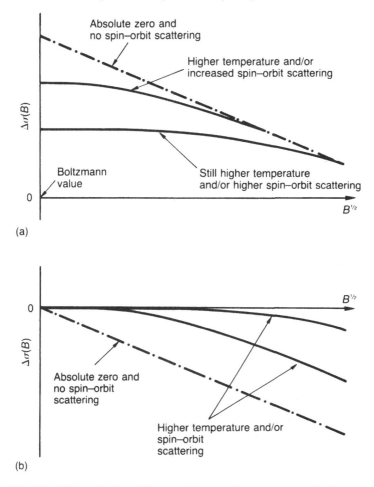

Fig. 14.1 The effect of spin–orbit scattering on the magnetoconductivity due to the enhanced interaction effect (schematic). (a) The change in conductivity due to the field B plotted against $B^{1/2}$ in relation to the Boltzmann value as zero. The dot-and-dash line is for $T = 0$ and no spin–orbit scattering. The two other curves show how $\Delta\sigma$ alters as either the temperature is raised; or spin–orbit scattering is increased; or both happen together. The two effects are exactly similar. (b) The changes in conductivity plotted from a common origin at which $\Delta\sigma(B)$ is zero at $B = 0$. The curves, apart from the physically unrealistic line for $T = 0$ and no spin–orbit scattering, all start off as B^2 and at high enough fields become parallel to the zero-temperature and zero spin–orbit line.

The magnetoconductivity, however, is not measured from the Boltzmann value but from the value of $\Delta\sigma$ for $B = 0$ at the relevant temperature. The curves are replotted with this new origin in Figure

14.1(b), which shows that as the temperature rises, the magnetoconductivity approaches its asymptotic form at higher and higher fields. Another way of expressing this is that the asymptotic form for high fields intersects the zero-conductivity line at approximately $g\mu B \simeq k_B T$. This follows from the slope of the asymptote, given in equation (14.3) for $\Delta\sigma$ versus $B^{1/2}$, and the shift of origin with temperature given in equation (13.5).

If now we take the exact theoretical results, we find that the magnetoconductivity due to spin-splitting at high fields $(g\mu B \gg k_B T)$ is:

$$\Delta\sigma(B) = (e^2/4\pi^2\hbar)\tilde{F}(kT/2\hbar D)^{1/2}[(g\mu B/k_B T)^{1/2} - 1.294] \qquad (14.5)$$

At low fields we have:

$$\Delta\sigma(B) = (e^2/4\pi^2\hbar)\tilde{F}0.056(g\mu B/k_B T)^2 \qquad (14.6)$$

where $\tilde{F}$ is defined in equation (A4.4).

We saw earlier that if $F \ll 1$, $\tilde{F} \simeq F$. A further point is that if $\tilde{F}$ and therefore F are zero, then according to equations (14.5) and (14.6) $\Delta\sigma(B)$ vanishes. This confirms that the exchange terms in the particle–hole channel make no contribution to the magnetoresistance. Equation (14.5) shows that the high-field asymptote intersects the axis at $(g\mu B/k_B T)^{1/2} = 1.294$.

This is a bare outline of the interaction effect insofar as the Zeeman splitting alters the contribution from the Hartree terms. There are, however, additional effects and contributions which we must look at before we attempt any comparison with experiment.

14.3 Spin–orbit scattering and magnetoresistance

We have discussed qualitatively how spin–orbit scattering alters the conductivity due to the interaction effect and its temperature dependence. We now apply a similar method to the magnetoconductivity, to which we have only one major contribution, that from the Hartree terms, which produce a positive magnetoresistance through the Zeeman splitting of the spin states.

Suppose now that the coherence time of an electron pair is limited not only by the temperature but also by the random effects of spin–orbit scattering. As we have already discussed in section 13.3, when both temperature and spin–orbit scattering act together, the probabilities of the two types of scattering add and the total probability becomes:

$$1/\tau_{\text{total}} = 1/\tau_T + \alpha'/\tau_{\text{so}} \qquad (14.7)$$

where $1/\tau_T = k_B T/\hbar$ and α' is a suitable weighting factor of order unity. Thus the net effect of spin–orbit scattering is, according to these arguments, to increase the effective temperature as seen by the magnetic field to T_{eff} given by:

$$T_{\text{eff}} = T + \alpha' \hbar / k_B \tau_{\text{so}} \qquad (14.8)$$

In the absence of spin–orbit scattering, we know that $\Delta\sigma(B)$ first decreases as B^2 and at higher fields it approaches the asymptotic line (see preceding section) and finally varies almost as $B^{1/2}$ as shown schematically in Figure 14.1(a).

When spin–orbit scattering is turned on, it raises the effective temperature and thereby reduces the magnetoresistance until, when $(\hbar/2g\mu B) \ll \tau_{\text{so}}$, $\Delta\sigma$ takes on the value it would have without spin–orbit scattering. The effects for several strengths of spin–orbit scattering are shown schematically in Figure 14.1(a). Identical effects could be produced by temperature. To complete the picture, in Figure 14.1(b) we plot $\Delta\sigma(B)$ as a function of B from a common origin at $B = 0$. This shows that, as the spin–orbit scattering increases, the magnetoconductivity decreases until ultimately there is no magnetoconductivity within the limited range of the magnetic fields employed. In section 16.4, these predictions are looked at in the light of experiment.

14.4 The Cooper channel

In the Cooper channel the two electrons execute the same path but in opposite senses and in the process one electron can still make use of the charge hologram of the other to recover its initial phase. Then, as in weak localisation, a magnetic field will induce a phase difference in the two wavefunctions from the magnetic flux that passes through the common orbit since they now execute it in opposite senses. This is called the orbital effect, as opposed to the effect of Zeeman splitting, and it makes a small contribution to the magnetoresistance.

The exchange terms make no Zeeman contribution to the magnetoresistance but the exchange terms in the Cooper channel contribute in a small way through the orbital effect.

The various possible contributions are summarised in Table 14.1.

Table 14.1 *Magnetoresistance contributions in the interaction effect.*

	Spin combination	Contribution
Exchange terms		
particle–hole	parallel	none
Cooper channel	parallel	orbital
Hartree terms		
particle–hole	parallel	none
	antiparallel	Zeeman
Cooper channel	parallel	orbital
	antiparallel	Zeeman and orbital

14.5 Zeeman splitting in weak localisation

The importance of Zeeman splitting in the magnetoresistance due to the interaction effect suggests that such splitting can also cause the dephasing of partial waves of opposite spin in weak localisation. This is indeed true. Let us therefore estimate the relative size of the two contributions, the orbital part discussed in Chapter 11 and the Zeeman splitting discussed here.

Let us define the critical field for the orbital contribution in weak localisation as that magnetic field whose dephasing time just equals that from inelastic scattering. This field (from equation (11.12)) is:

$$B_{\mathrm{OWL}} = \hbar/4eD\tau_{\mathrm{in}} \qquad (14.9)$$

The comparable field for the Zeeman splitting B_{ZWL} is the field that just dephases the two partial waves in the same time τ_{in}. The rate of change of phase between the two waves in such a field is $\omega = g\mu B_{\mathrm{ZWL}}/\hbar$ and so the critical field occurs when $\omega \simeq 1/\tau_{\mathrm{in}}$. This gives:

$$B_{\mathrm{ZWL}} = \hbar/g\mu\tau_{\mathrm{in}} \qquad (14.10)$$

The ratio of the two critical fields in weak localisation is thus:

$$B_{\mathrm{OWL}}/B_{\mathrm{ZWL}} = g\mu/4eD \qquad (14.11)$$

If we put the Bohr magneton $\mu = e\hbar/2m$, $g = 2$ and $D = v_{\mathrm{F}}l/3$, we find:

$$B_{\mathrm{OWL}}/B_{\mathrm{ZWL}} \sim h/mv_{\mathrm{F}}\tau_0 \sim 1/k_{\mathrm{F}}l \sim h/e_{\mathrm{F}}\tau_0 \qquad (14.12)$$

Normally in Boltzmann-type metals we expect $E_{\mathrm{F}} \gg h/\tau_0$; this will also be true of metallic glasses made from non-transition metals so that in these the orbital field will be very small compared to the Zeeman field.

Generally therefore in weak localisation the Zeeman contribution is small compared to the orbital contribution to the magnetoresistance. For some of the alloys that we discuss, however, $h/E_F\tau_0$ may approach unity so that the two effects may be comparable and for them the full formulae for the magnetoconductivity due to weak localisation must include the Zeeman contribution.

14.6 The Maki–Thompson correction

This correction (Maki 1968, Thompson 1970) came originally from a study of superconducting fluctuations, from which it derives. It can, however, contribute at temperatures far from the superconducting transition temperature. It arises from electron pairs in the Cooper channel, which thus execute closed paths in opposite senses. These electron pairs resemble the partial waves in weak localisation except that they require a coupling mechanism. Indeed the resulting change in conductivity is essentially just the singlet contribution in weak localisation modified by a temperature-dependent coupling parameter. The correction is important in magnetoresistance and often has to be included there. Some details are given in Appendix A5.

14.7 Electron–electron scattering in disordered metals

There is another effect of the enhanced electron interaction in a disordered metal that is important, not for the interaction effect itself, but because it has possible consequences for weak localisation. This is the scattering of electrons by each other; even if this causes no significant resistance, it can cause dephasing and so alter the temperature dependence of the resistance.

In section 7.1 we saw that electron–electron collisions varied with temperature as T^2 in ordered metallic materials. We can now show how this temperature dependence is altered in a disordered metal. We saw in Chapter 12 and Appendix A3 that the self-energy of the interacting electrons was complex and that the imaginary part gave rise to a finite lifetime of the interacting state. So we can use the imaginary part of the self-energy to find the lifetime of an electron in diffusive motion interacting with another and thus get the probability of electron–electron scattering in a disordered metal. This is given in Appendix A6 and shows that now the scattering varies with temperature as $T^{3/2}$.

14.8 Diagrammatic techniques

Some of the effects of disorder are so strange and subtle that you may wonder if there are still more elaborate modes of interaction yet to be found. After all, if electrons can do all these weird things, why can't they do still weirder ones?

The answer is that they may, but there are two points to bear in mind. First, diagrammatic techniques in perturbation theory make it possible to examine systematically the different possible modes of interaction and identify those that are likely to be significant. It was through such techniques that these effects were first discovered. Second, experiments can help to guide the theory and show up any major gaps that may exist. The effects we are discussing show up even more dramatically in low-dimensional systems and their two-dimensional counterparts have been examined experimentally with considerable thoroughness and success in thin films.

A final point is that diagrammatic techniques do not reveal the physical nature of the interactions and the physical interpretation of the theoretical findings can take time. Bergmann (1983, 1987) in a series of papers has played an important and valuable role in this process and I think most of us would accept his arguments that not only does such interpretation make the subject clearer and may even show up errors or incompleteness but it often reveals unsuspected links with other parts of physics.

14.9 The Hall effect

We have discussed the Hall effect in the context of its 'anomalous' sign in many transition metal alloys. Now we must look at how the enhanced interaction effect alters it.

We can write the Hall coefficient in terms of the conductivity components σ_{xx} and σ_{xy}. Here we envisage the magnetic field B applied in the z-direction with the current and Hall field in the xy-plane. The Hall coefficient R_H can then be written:

$$R_H = \sigma_{xy}/B(\sigma_{xx})^2 \qquad (14.13)$$

Theory indicates that weak localisation causes no change in R_H but the enhanced interaction effect, while leaving σ_{xy} unchanged, causes R_H to change in response to the change $\Delta\sigma_{int}$ induced in σ_{xx}, i.e. the ordinary conductivity σ. If $\Delta\sigma_{int}$ represents the change in conductivity due to the enhanced interaction effect we have:

$$\Delta R_H/R_H = -2\Delta\sigma_{int}/\sigma = 2\Delta\rho_{int}/\rho \qquad (14.14)$$

This relationship shows that the Hall coefficient becomes slightly temperature dependent because, as we saw in Chapter 13, $\Delta\sigma_{int}$ changes with temperature although, as we saw there, such changes are small. Indeed the early measurements of the Hall effect in metallic glasses suggested that R_H was independent of temperature and it needed the subsequent more accurate measurements to reveal a temperature dependence, which in fact tallies with equation (14.14).

$\Delta\sigma_{int}$ varies as $+T^{1/2}$ and so equation (14.14) predicts that, if R_H is positive, the Hall coefficient should decrease with this power law; if, on the other hand, R_H is negative, it should increase with this power law.

Figure 14.2 shows the experimental values of the Hall coefficient of six metallic glasses together with the $T,^{1/2}$ dependence which is the least-square-fit to the points. The $Ni_{64}Zr_{36}$ alloy has a negative Hall coefficient; the others are positive. Moreover there is no measurable temperature variation in the resistivity of the relatively low-resistance alloy $Pd_{80}Si_{20}$. The curves fit the data qualitatively and there is order of magnitude agreement with the resistivity data. Later, more accurate measurements by Drewery and Friend (1987) on Cu–Ti amorphous thin films give quantitative agreement with the theory (Figure 14.3).

There seems little doubt that the temperature dependence of R_H is due to the enhanced interaction effect and potentially this dependence, which discriminates between weak localisation and the interaction effect, is most valuable. Unfortunately the change with temperature is small and, in typical glasses, the Hall effect itself is small. Moreover any tiny amounts of crystallinity in the samples can cause large errors. Since specimen perfection is difficult to achieve, all these factors make it very hard to measure R_H with the necessary high accuracy to exploit its potential usefulness.

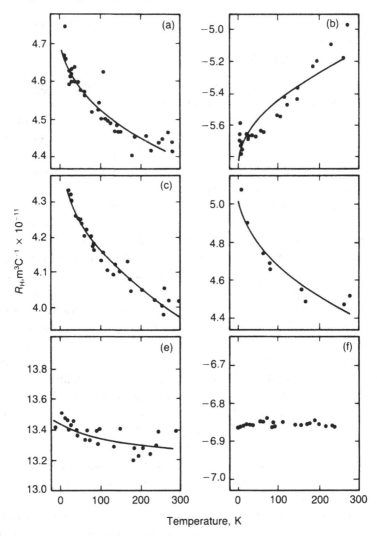

Fig. 14.2 The variation with temperature of the Hall coefficient in metallic glasses due to the interaction effect. (a) $Pd_{30}Zr_{70}$; (b) $Ni_{64}Zr_{36}$; (c) $Ni_{24}Zr_{76}$; (d) $Fe_{24}Zr_{76}$; (e) $Cu_{50}Ti_{50}$; (f) $Pd_{80}Zr_{20}$. The solid lines show $T^{1/2}$ variation fitted to the data. (Experimental data from Schulte *et al.* 1984; theoretical interpretation and diagram after Gallagher *et al.* 1984.)

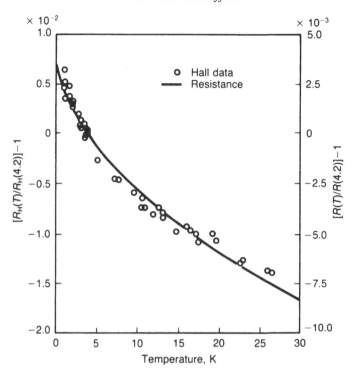

Fig. 14.3 The variation with temperature of the Hall coefficient of a Cu–Ti alloy compared with the variation to be expected from the resistivity due to the enhanced electron interaction. (After Drewery and Friend 1987.)

15

The thermopower of metals and alloys

15.1 Definitions

The origin of the thermoelectric effects is very simple. They arise because an electric current in a conductor carries not only charge but also heat. Consequently when an electric current flows through the junction of one conductor with another, although the charge flow is exactly matched, there is in general a mismatch in the associated heat flow; the difference is made manifest as the Peltier heat. If the current flows through a conductor in which there is a temperature gradient the heat shows up as the Thomson heat which is the heat that must be added to or subtracted from the conductor to maintain the temperature gradient unchanged; the electric current behaves as if it were a fluid with a heat capacity (either positive or negative). The third manifestation of thermoelectricity is the Seebeck effect which is the inverse of the other two. In this a heat current is established by means of a temperature gradient and this produces an electric current. However this cannot be done with a single material since in such a closed circuit the current induced in one part would cancel that in the other. Instead two materials are needed; moreover it is more convenient to measure not the circulating current that results but the emf that arises when the electrical circuit is broken. More explicitly, if conductor A is connected to conductor B at its two ends and the two junctions are maintained at different temperatures, an emf appears in the circuit.

All these effects are related quantitatively to each other by the Thomson or Kelvin quasi-thermodynamic relations. The definitions and relationships can be summarised as follows: if in the Seebeck effect a temperature difference ΔT between the junctions of the conductors A and B produces a voltage difference ΔV (see Figure 15.1) the Seebeck coefficient S_{AB} is defined as:

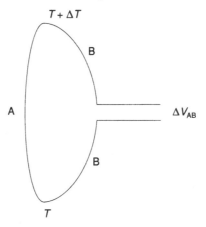

Fig. 15.1 The Seebeck effect.

$$S_{AB} = \Delta V/\Delta T \qquad (15.1)$$

in the limit as ΔT tends to zero.

If the two conductors A and B are joined at an isothermal junction and a charge q flows across the junction the Peltier heat Π_{AB} is defined as:

$$\Pi_{AB} = Q/q \qquad (15.2)$$

where Q is the heat reversibly given out or absorbed at the junction. If q is positive and heat is reversibly given out the Peltier heat is defined as positive and A is positive with respect to B.

The Thomson coefficient μ_A of conductor A measures the heat absorbed (or given out) reversibly when unit positive charge passes through unit positive temperature difference in the conductor. It is defined as positive if this heat is absorbed.

The Thomson coefficient thus characterises a single material; it turns out that the Peltier and Seebeck coefficients can themselves be split up so as to refer to individual materials. Thus we write:

$$\Pi_{AB} = \Pi_A - \Pi_B \qquad (15.3)$$

and

$$S_{AB} = S_A - S_B \qquad (15.4)$$

where Π_A, S_A etc. are characteristic of materials A or B alone. S_A or S_B is then referred to as the absolute thermoelectric power of the particular material.

The Thomson (Kelvin) relations state that:

$$S = \Pi/T \qquad (15.5)$$

and

$$S = \int_0^T (\mu/T)\mathrm{d}T \qquad (15.6)$$

Here T is the absolute temperature. Relation (15.5) is analogous to the equilibrium thermodynamic relation between the entropy (analogous to thermopower) and latent heat (analogous to Π_{AB}). Relation (15.6) is analogous to that between entropy and specific heat and indicates how the absolute thermopower of a metal could be measured; it turns out however that, because superconducting electrons carry zero entropy, superconductors have zero thermopower so that in the temperature range where these exist the absolute thermopower of a conducting material can be determined directly by measurements on a thermocouple of which the other material is a superconductor.

Although I have referred to the 'heat' carried by the electric current the reader may be suspicious of this usage since in learning thermodynamics, we are repeatedly warned that we can only recognise heat in what we may call 'external transactions'. This is a perfectly proper objection and we ought more strictly to confine our attention to entropy. Since, however, some people are frightened by this word it seems worthwhile to use the more comfortable and familiar notion of heat provided that we exercise due care in defining what we mean by 'heat' in this context. The foregoing description gives correctly the essential ideas and we turn now to the actual calculation of the thermopower of a metal.

15.2 The calculation of thermopower in a metal

Conceptually the easiest thermoelectric effect is the Peltier effect; the Peltier heat of a metal is just the heat per unit charge carried by an electric current through the metal. No temperature gradients are involved which helps to simplify the calculations. If we can calculate the Peltier coefficient we can then use equation (15.5) to derive the thermopower, which is the quantity most commonly measured and considered.

If we establish an electric current in a metal by the application of an electric field, the Peltier coefficient can be written:

$$\Pi = \text{heat current/charge current.}$$

If the electric field is ϵ_x in the x-direction, we know from equation (3.31) that the charge current j_x is given by:

$$j_x = -(e^2\epsilon_x/4\pi^3\hbar)\int\int \tau(v_x^2/v)\mathrm{d}S_E(\mathrm{d}f_0/\mathrm{d}E)\mathrm{d}E \qquad (15.7)$$

where the first integral is over a surface S_E of constant energy E and the second over all energies.

Let us now introduce the partial conductivity for electrons of energy E defined as:

$$\sigma_x(E) = (e^2/4\pi^3\hbar)\int \tau(v_x^2/v)\mathrm{d}S_E \qquad (15.8)$$

Then j_x can be written:

$$j_x = -\epsilon_x\int \sigma_x(E)(\mathrm{d}f_0/\mathrm{d}E)\mathrm{d}E \qquad (15.9)$$

In this $\sigma_x(E)$ is the conductivity of those conduction electrons that have energy E; we confine ourselves to elastic scattering so that electrons on a given energy shell stay on it even after scattering. We thus know the charge current as the integral over the different energy shells and the limitation to elastic scattering is not too severe in practice. Such scattering includes scattering by chemical impurities, by many physical defects and, perhaps surprisingly, by phonons at temperatures above about one-third of the Debye temperature. This last type of scattering, while not truly elastic, is effectively so.

Having calculated the charge current, our next task is to calculate the corresponding heat current. To do this we must first define the heat carried by an electron of given energy in a degenerate Fermi gas. Let the Fermi energy of the gas be E_F, its pressure p and its volume per electron V. Now consider an electron whose energy is E and entropy s (this is the entropy per electron associated with all the electrons of energy E). The chemical potential of such electrons is the same as their Fermi energy:

$$E_F = E - Ts + pV \qquad (15.10)$$

and at constant temperature the chemical potential of the electron gas must be everywhere the same, i.e. it is common to the two metals at a junction.

We take for the heat carried by this electron the quantity $E - E_F$. This is a basic tenet of the thermodynamics of irreversible processes. Since by equation (15.10) it is equal to $Ts - pV$ it can be interpreted as saying that the heat associated with the electron is determined by its entropy s and the absolute temperature and that the pV term takes into account the work done if, for example, the gas moves into a region where its density is changed.

If we take the expression (15.9) for the charge current, we can convert it into an expression for the corresponding heat current as follows. Dividing equation (15.9) by e, the electronic charge, we get an integral over the number of electrons flowing in unit time at each energy. If therefore we multiply the number at each energy by the energy carried by such electrons, i.e. by $(E - E_F)$, we then get the heat current for that energy. For the total heat current we then integrate over all energies. The heat current is thus:

$$\int_{-\infty}^{+\infty} [\sigma_x(E)(E - E_F)(df_0/dE)dE]/e \qquad (15.11a)$$

the charge current is:

$$\int_{-\infty}^{+\infty} \sigma_x(df_0/dE)dE \qquad (15.11b)$$

and the ratio of heat current to charge current becomes:

$$\Pi = \left[\int_{-\infty}^{+\infty} \sigma_x(E)(E - E_F)(df_0/dE)dE \right] \Big/ e\left[\int_{-\infty}^{+\infty} \sigma_x(df_0/dE)dE \right]$$
$$(15.11c)$$

Since in cubic or amorphous metals σ is a scalar quantity we can omit the suffix x.

This expression can already give us useful information. The denominator is just $-\sigma e$, the negative of the electrical conductivity times the electronic charge. The numerator is best considered as a function not of E but of $\epsilon = E - E_F$. We can then write (cf. 3.14):

$$f_0 = 1/[1 + \exp(\epsilon/k_B T)]$$

so that

$$df_0/dE = df_0/d\epsilon = -\exp(\epsilon/k_B T)/k_B T[1 + \exp(\epsilon/k_B T)]^2 \qquad (15.12)$$

which is always negative and completely symmetrical about the origin of ϵ. Moreover the new limits of integration are $+\infty$ and $-E_F$; the latter is so large compared to $k_B T$ that the integrand is completely negligible at this and more negative values of ϵ. Thus we can write the lower limit as $-\infty$. If therefore $\sigma(E)$ is constant, independent of E, the numerator vanishes because the factor $(E - E_F) = \epsilon$ ensures that all contributions to the integral at energies below E_F are exactly cancelled by the corresponding contributions above E_F. This therefore tells us that unless the partial conductivity $\sigma(E)$ varies with energy the thermopower of the metal that arises from elastic scattering is zero.

If, as is usually the case, there is an energy dependence of $\sigma(E)$ we can take account of it and use equation (15.11c) to derive an expression for the thermopower. We take only the first-order dependence on energy and write:

$$\sigma(E) = \sigma(E_F) + (\partial\sigma/\partial E)(E - E_F) \qquad (15.13)$$

where the derivative is understood to be evaluated at E_F. If we put this in the numerator of equation (15.11c) the first term on the right of equation (15.13) contributes nothing because it is constant. If for convenience we put

$$(E - E_F)/k_B T = \epsilon/k_B T = u \qquad (15.14)$$

the limits on u are from $-\infty$ to $+\infty$. The numerator of equation (15.11c) then becomes:

$$(\partial\sigma/\partial E)(k_B T)^2 \int_{-\infty}^{+\infty} u^2 (df_0/du)du \qquad (15.15)$$

so that finally, putting in the value of the definite integral in equation (15.15), which is $-\pi^2/3$, we get for the Peltier heat:

$$\Pi = (\pi k_B T)^2 (\partial\sigma/\partial E)/3e\sigma \qquad (15.16)$$

This in turn yields the Mott formula for the thermopower:

$$S = \pi^2 k_B^2 T(\partial \log \sigma/\partial E)/3e \qquad (15.17)$$

This shows that the thermopower arising from elastic scattering is directly proportional to the absolute temperature. Its sign depends on the sign of the charge e of the carriers and of the derivative of $\sigma(E)$ with respect to energy at the Fermi level.

It is as if the conductivity, or rather its change with energy, acted as an energy filter. If it discriminates in favour of the electrons of high energy (above the Fermi level) and lets through more of them than those below the Fermi level, the sign of $\partial(\log \sigma)/\partial E$ is positive and the sign of S is that of the charge carriers. If, on the other hand, it favours the low energy electrons, the sign of S is opposite to that of the charge carriers. Finally if it is undiscriminating and lets through all electrons equally then there is no thermopower at all.

We saw earlier that the thermopower is simply the entropy per unit charge of the charge carriers. The fact that it is proportional to the absolute temperature reflects the fact that the entropy (and specific heat) of a degenerate electron gas behaves in just this way.

The Mott formula, equation (15.17) is, apart from the restriction to elastic scattering, of very general application in metals or alloys; it can apply to crystals, liquids or glasses. We shall find it of great value in our discussions of the thermopower of metallic glasses but I should emphasise that only the temperature dependence of S is readily predicted by the theory; the calculation of $\partial\sigma/\partial E$ may be beyond the power of present calculations.

This is however not the whole story. The heat or entropy that contributes to the Peltier or Seebeck effects has so far been ascribed to the electron gas alone. However we have already seen that an electric current can carry with it phonons and these phonons will then contribute to the heat or phonon drag component; it simply adds to the other intrinsic or 'diffusive' component. In metallic glasses however the disorder is so great that for the most part it suppresses the phonon drag. For this reason we shall not consider it further here.

15.3 The thermopower of metallic glasses

Let us now see how this theory applies to metallic glasses. Since the electrical conductivity is largely limited to elastic scattering of the conduction electrons from the disorder in the metallic glass, the thermopower should be given by the Mott formula, equation (15.17). The high disorder will, as we saw, 'kill off' any phonon drag component and so we should expect to find that the thermopower of these glasses is directly proportional to the absolute temperature i.e. $S \propto T$. This expectation is securely based on equation (15.17), which is of wide generality and the few conditions that limit it are well satisfied by the metallic glasses. To test the prediction of the formula, it is convenient, instead of the obvious plot of S versus T, to plot S/T versus T, which should, according to equation (15.17), yield a constant independent of T. The results of experiment on some metallic glasses are shown as a plot of S versus T in Figure 15.2(a) and as S/T versus T in Figure 15.2(b). These demonstrate unequivocally that the simple theory is inadequate.

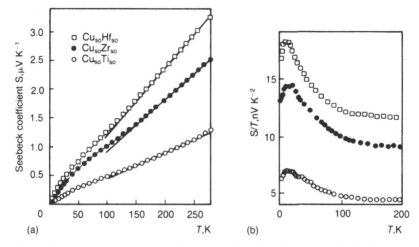

Fig. 15.2 Thermopower of metallic glasses $Cu_{50}Ti_{50}$, $Cu_{50}Zr_{50}$ and $Cu_{50}Hf_{50}$. (a) S versus T. (b) S/T versus T. (After Gallagher and Hickey 1985.) At temperatures above about 150 K, S is directly proportional to T as expected by conventional theory. At lower temperatures there are departures from proportionality. These departures are seen in the plot of S/T versus T and show that the thermopower is enhanced at low temperatures.

The source of the discrepancy lies in some many-body effects that have been neglected in our theoretical account, although, as we shall see, the same effect has already been seen under different circumstances in crystalline metals.

As Figure 15.2 illustrates for CuTi, CuZr and CuHf, the thermopower of many metallic glasses does not vary in direct proportion to the absolute temperature as we would expect from equation (15.17). There is approximate proportionality to T at high temperatures (above about 150 K) and again, with a larger constant of proportionality, at low temperatures. How can we account for this temperature variation? The answer lies in an effect already well known in crystalline metals, to which we have already alluded, namely the interaction of the conduction electrons with the ions. As we discussed earlier, the theory of superconductivity devised by Bardeen, Cooper and Schrieffer (the BCS theory) is based on this interaction; it gives rise to an attraction between electrons near the Fermi level and thereby ultimately to the superconducting state. Even in those metals that do not show superconductivity, the interaction shows up as an enhancement of the electronic heat capacity and the associated entropy. (This additional entropy of the Fermi electrons can be thought of as arising from the interaction with the ions at low temperatures; the electrons acquire a greater effective mass and so a higher density of states.)

It is this last feature that gives the clue to what is happening to the thermopower. I have described the thermopower of a conductor as the entropy per unit charge carried by the current. If this description is correct, then any phenomenon that alters the entropy of these current carriers should alter the thermopower. This intuitive argument is correct and allows us to understand the main features of the thermopower of metallic glasses.

We saw earlier that if ω_{max} is the highest frequency of oscillation sustainable by the ions, only electrons within an energy range of $\hbar\omega_{max}$ of the Fermi level can participate in the interaction. If for convenience we use a Debye model for the ionic vibrations, we can use the approximation that $\hbar\omega_{max} = k_B\theta_D$.

The dispersion curve of the electrons around the Fermi level is as shown in Figure 15.3, which is valid for low temperatures. The slope of the curve at the Fermi level is reduced, implying that the density of states is enhanced by the electron–phonon interaction. This also means that the entropy at a given low temperature is correspondingly enhanced. At high temperatures, however, (T about θ_D) the electrons shake off the interaction with the ions, the dispersion curve reverts to its unenhanced shape and the entropy is correspondingly reduced to the value it would have without the interaction. As we have already noted the effect can be seen in the electronic heat capacity of the alkali metals. The measured values are significantly larger than the theoretical values calculated from the band structure of the metals. The specific-heat enhancement is also to be expected in metallic glasses but it cannot be inferred from the measured electronic heat capacity because we cannot accurately establish the unenhanced value. It should, however, show up in the thermopower because this depends on the entropy associated with the Fermi electrons and this, in turn, depends on the density of states at the Fermi level. The anomalous temperature dependence of the thermopower already mentioned is ascribed to this effect.

What happens is as follows. At low temperatures the entropy of the Fermi electrons is fully enhanced and this shows itself as an enhancement of the thermopower. As the temperature is raised the enhancement is reduced by thermal excitations until at high temperatures it is completely destroyed and the unenhanced thermopower is revealed. This explains why the thermopower is approximately proportional to T at low and at high temperatures but with different coefficients. The ratio of the two

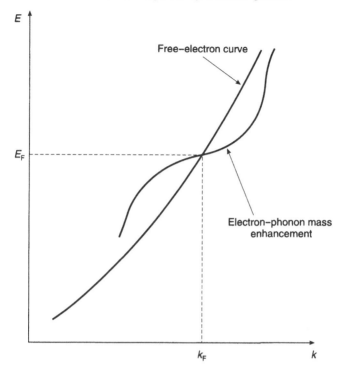

Fig. 15.3 Dispersion curve of electrons illustrating the change of slope due to electron–phonon interaction within about $\pm k\theta_B$ of the Fermi level.

coefficients should therefore be a measure of the full enhancement factor. Inasmuch as the enhancement of the thermopower is the same as that involved in the heat capacities (which is unlikely to be strictly true though a useful first approximation) we can test the theory by comparing the two. In those glasses that show superconductivity a measure of the enhancement can be derived from their superconducting properties. This is expressed through the parameter λ_{ep} discussed in section 7.4. If this can be found from the superconducting properties, we can use it to compare with the thermopower results. The ratio of the fully enhanced to the unenhanced heat capacity is $1 + \lambda_{ep}$ and this should be approximately the same for the thermopower. A comparison is made for several metallic glasses in Table 15.1. The agreement is convincing evidence that we have indeed found the source of the 'anomalous' temperature dependence.

As well as the magnitude of the effect, we can also compare the temperature dependence of the thermopower in different alloys. This is expected and indeed found to scale with T/θ_D.

Table 15.1 *Comparison between enhancement factors λ_{ep} (at 0 K) derived from thermopower and from superconductivity*

Metallic glass	$1 + \lambda_{ep}$ (Thermopower)	$\dfrac{1 + \lambda_{ep}}{\text{(Superconductivity)}}$
$Cu_{50}Ti_{50}$	1.4	1.33
$Cu_{50}Zr_{50}$	1.5	1.43
$Cu_{50}Hf_{50}$	1.5	1.40
$La_{76}Al_{24}$	2.0[a]	1.8
$La_{78}Ga_{22}$	2.0[a]	1.8

[a] Negative thermopower; the others are positive.

The treatment so far gives the essential physics of the thermopower of these metallic glasses but a more complete treatment recognises that the enhancement is a function of the energy of the electrons (just as the partial conductivity is) and this must be put into the calculation. Likewise the calculation has to be made as a function of the temperature. Such calculations have been made and compared with more detailed experimental values of the thermopower at low temperatures. These are show in Figure 15.4[1].

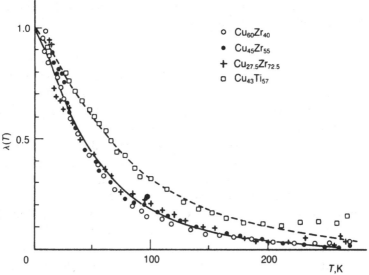

Fig. 15.4 The electron–phonon enhancement factor versus temperature as derived from thermopower measurements (Gallagher 1981). Solid lines are theoretical fits and the temperature is scaled according to the Debye temperature. (After Kaiser 1982.)

The conclusion is that the thermopower shows very directly the influence of the phonon-induced interaction between conduction electrons; indeed if we measure $S(T)$ and plot $(S/T) - (S/T)_{\text{high temp.}}$ as a function of temperature, we have the simplest experimental way of showing (at least approximately) how λ_{ep} changes with temperature.

16

Comparison with experiment

16.1 Introduction

We have now seen in some detail how weak localisation and the interaction effect can modify the electron transport properties of electrons in metallic glasses or, more specifically, of electrons that are subject to strong elastic scattering, whether this be in the crystalline or the amorphous phase. What this survey shows is that many of the qualitative features to be expected are indeed observed in the resistivity, magnetoresistance and Hall coefficient of metallic glasses. The final question is: how far do the theories provide a quantitative account of the experiments?

It is at once clear, I think, why it is difficult to answer this question unequivocally. There are so many parameters that can influence the behaviour of these properties that unless some can be controlled or eliminated there are too many adjustable quantities to make possible a convincing comparison between theory and experiment.

One common way to overcome this problem is to make measurements of a range of properties so that a given specimen is very well characterised and as few as possible of the relevant parameters are left undetermined. So let us decide what quantities we know or can deduce with some reliability from experiment.

We can measure the low-temperature heat capacity of the metallic glass to find the term linear in temperature, which allows us to deduce the density of states at the Fermi level. In order to interpret the thermopower we would like to know the electron–phonon enhancement factor in the alloy; if it is a superconductor we can derive this from our knowledge of its superconducting properties.

From the density of states, the measured conductivity and the Einstein relation we can deduce the electron diffusion coefficient D. For this we must use the fully enhanced value of the density of states; the physical argument for this is that the electrons are subject to all the many-body effects present in the metal and must respond to them in their diffusion. The view that we should use the bare density of states probably comes from the mistaken belief that this is always appropriate when discussing electron transport.

What we would next like to know is the number of conduction electrons per unit volume. In simple metals the composition of the alloy and the valence of the constituents should give us the answer. That means that, with a knowledge of the molar volume, we can calculate the Fermi radius k_F and, insofar as the Fermi surface is a valid concept, this radius is reliable, depending only on the volume in k-space required to house the requisite number of electron states. From this and the density of states we can also infer the Fermi velocity.

The Hall coefficient can be used to test whether the glass is behaving like a single band, free-electron metal and if so we can then calculate the Thomas–Fermi screening radius and thence the value of the interaction parameter F.

With transition metal alloys there is no simple, direct way of getting information about the radii of the two spherical parts of the Fermi surface. The Hall coefficient may in some circumstances give us an estimate of the more mobile hybridised s–d electrons and hence an estimate of the number of the d-like electrons. But in general there are severe uncertainties about the band structure of these alloys.

16.2 Conductivity of Cu–Ti metallic glasses

Let us start our experimental comparison with some measurements on a range of Cu–Ti glasses at low temperatures made by Schulte and Fritsch (1986). The problem is to disentangle the weak localisation from the interaction effect. In evaluating the weak localisation contribution, spin–orbit scattering has to be included and the first step is to analyse the low-field magnetoresistance, quadratic in B. In the interaction effect only the particle–hole channel with Zeeman splitting is needed since the Cooper channel and the orbital terms are very small. The authors derive the value of D as indicated above and take the value of $\tilde{F}$ from earlier work on the Hall coefficient. This enables them to determine the strength of the interaction contribution to the low-field magnetoresistance and so, by subtraction from the measured total, to find the part due to weak localisation. An analysis of this part gives (a) the spin–orbit scattering time, which is independent of temperature; and (b) the inelastic scattering or, more generally, the dephasing time together with its temperature dependence.

Having analysed the low-field magnetoresistance, the authors then look at how the conductivity in zero field varies with temperature. Again there are the two contributions: one from weak localisation and one from the interaction effect. The two contributions behave quite differently and this is illustrated very clearly in Figure 16.1,

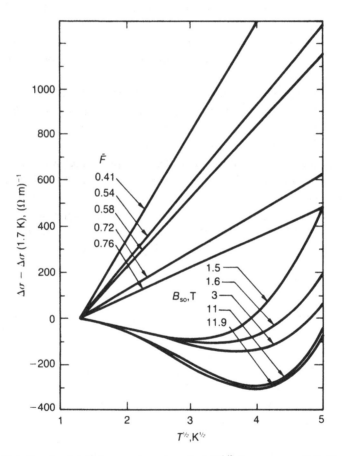

Fig. 16.1 Conductivity change plotted against $T^{1/2}$ for a range of CuTi metallic glasses (the copper concentration increases from the top curve downwards). The upper straight lines show the theoretical contribution from the Coulomb interaction, with the different values of $\tilde{F}$ as shown, fitted to the experimental data. The lower curves are the contributions from weak localisation with different degrees of spin–orbit scattering. B_{so} in tesla is defined as $B_{so} = \hbar/4eD\tau_{so}$. (After Schulte and Fritsch 1986.)

which is a plot of the conductivity changes versus $T^{1/2}$. The interaction part is seen as a series of straight lines, one for each of the different alloys. The lines are straight because the interaction contribution to the conductivity is assumed to vary as $T^{1/2}$ and the different slopes correspond to different values of $\tilde{F}$. The weak localisation part is shown in the lower curves which start off with a negative slope and then change to a positive one. This set again refers to the same alloys but now the spin–orbit scattering determines their different shapes. Finally, the two sets of curves are added together in pairs to generate the theoretical totals that are compared with experiment in Figure 16.2. The parameters $\tilde{F}$ and τ_{so} have here been chosen to optimise the fit in all cases but there is also one curve, shown as a dot-dash, which is calculated for $Cu_{44}Ti_{56}$ from the parameters derived from the magnetoresistance.

This analysis ignores the effect of spin–orbit scattering on the enhanced interaction effect which, as we discussed in Chapter 13, alters the temperature variation of the associated conductivity from the simple $T^{1/2}$ dependence. The positive curvature induced by the spin–orbit scattering is, at least qualitatively, in the right sense to improve the agreement in Figure 16.2.

Even without this additional correction, the overall impression is, I think, that the theory gives a good account of the experimental results. The temperature dependence of the conductivity is not simple and the theory explains it in a natural way, with reasonable values of the parameters and without having to invoke special conditions.

There have been several other investigations of Cu–Ti glasses and the data from a number of these are collected in Table 16.1. These serve to indicate that there are differences between the parameters derived by different observers, partly through different modes of analysis and partly through the use of different versions of the theory.

Table 16.1 shows that there is some consistency between the different observers; the commonest problem, however, is to reconcile the data from the magnetic field dependence, which in general fit the theory with reasonable values of the parameters, with the data derived from the temperature dependence in zero field, which not only tend to yield different but sometimes unphysical values of the associated parameters. Partly this is due to the neglect of spin–orbit scattering on the enhanced electron interaction on the conductivity but it may also be that the electronic structure of these transition metal alloys prevents a very close fit of theory and experiment.

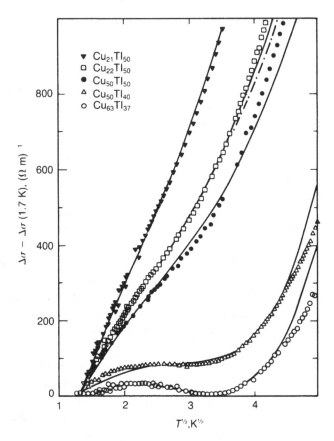

Fig. 16.2 Conductivity change as a function of $T^{1/2}$ for the alloys of the preceding figure. The points represent the experimental data and the continuous curves are derived by adding together the two contributions shown in the previous figure. The dot-dash curve is calculated from parameters derived from the magnetoconductivity of $Cu_{44}Ti_{56}$. (After Schulte and Fritsch 1986.)

Before we turn to an example of a rather simpler alloy system, let us look at a wider range of metallic glasses to gain a general impression of how well and how widely the theory can be applied.

16.3 Magnetoresistance in a wide range of glasses

16.3.1 Weak localisation

Bieri *et al.* (1986) reported measurements on a wide range of metallic glasses made by the technique of sputtering: $Mg_{80}Cu_{20}$, $Cu_{50}Lu_{50}$,

Table 16.1 Some data on Cu–Ti metallic glasses

Alloy	Resistivity $(\mu\Omega\,cm)$	D $(cm^2\,s^{-1})$	F	τ_{so} (ps)	p^a	Source
$Cu_{35}Ti_{65}$	182	0.24	0.40	7.3(T) 6.0(B)		(a)
	182	0.24	0.45	10.4	(2)	(b)
$Cu_{41}Ti_{59}$		(0.31)	0.41	3.5	2.53	(c)
$Cu_{42.5}Ti_{57.5}$		0.25	0.26	2.5	(2)	(b)
$Cu_{44}Ti_{56}$		(0.31)	0.54	2.9	2.64	(c)
$Cu_{50}Ti_{50}$		0.24	0.37	1.0	(2)	(b)
		(0.31)	0.58	1.8	2.78	(c)
				2.2		(d)
$Cu_{60}Ti_{40}$		(0.31)	0.72	0.5	2.99	(c)
	187	0.28	0–0.32	0.08–0.25	2.0–2.5	(e)
$Cu_{63}Ti_{37}$		(0.31)	0.76	0.4	2.84	(d)
$Cu_{65}Ti_{35}$	182	0.30				(e)
		0.31	0.45	0.2	(2)	(b)
				1.2		(d)

^{a} p is defined by the relation: $a/\tau_{in} \propto T^p$.
(a) Hickey et al. (1987) – the symbol (T) indicates a value derived from the temperature dependence of conductivity and (B) that from magnetic field dependence; (b) Hickey et al. (1986) – the value of $p = 2$ is assumed; (c) Schulte and Fritsch (1986) – the value of $D = 0.31$ was estimated for all the alloys; (d) Howson et al. (1986); (e) Lindqvist and Rapp (1988) – the values for $Cu_{60}Ti_{40}$ are just one of five possible fits that they give for their data.

$Pd_{80}Si_{20}$, $Cu_{50}Y_{50}$, $Cu_{57}Zr_{43}$, $Y_{80}Si_{20}$ and some ternary alloys. Their aim was to work in a range of temperatures and magnetic fields where weak-localisation effects were predominant in most of the alloys and to illustrate the importance of spin–orbit scattering on these effects. They estimated that if the magnetoresistance were due to the interaction effect, the $B^{1/2}$ dependence at 4.2 K would not be seen until B reached a value of about 10 T, whereas their measurements showed that the magnetoresistance varied as $B^{1/2}$ at less than 0.6 T in $Cu_{50}Y_{50}$ and at less than 0.2 T in $Cu_{57}Zr_{43}$. So they concluded that the magnetoresistance was due largely to weak localisation.

The general results of the experiments confirmed that the magnetoresistance does not depend on the field direction, that any Kohler contribution is negligible, that the coefficient of the B^2 term from low-field measurements depends on temperature and that of the $B^{1/2}$ term at high fields does not. We have already discussed the impressive qualitative features of some of these experiments in Chapter 11 and now we turn to the quantitative comparison between theory and experiment.

The magnetoresistance of the first four alloys mentioned above was analysed in terms of weak localisation only. First the high-field measurements were used to find how closely the data agreed with the theoretical predictions. Since there were small discrepancies in the size of the $B^{1/2}$ term, these were absorbed in a factor α placed in front of equations (A1.4) and (A1.5). Values of α for various alloys are given in Table 16.2. They range in value between 1 and 1.5.

Then the low-field data were analysed to determine τ_{in} as a function of temperature. By fitting the full field dependence of the magnetoresistance values of the spin–orbit lifetime could be found; these are most accurately determined where the maximum in the magnetoresistance occurs within the field range of the experiments.

Figure 16.3 shows the experimental results and theoretical fits for $Mg_{80}Cu_{20}$; the parameters used in fitting the curves are shown in Table 16.2. Values of D were derived from the conductivity and the free-electron density of states. If there were no spin–orbit scattering, the magnetoresistance would be negative at all fields but the fact that there is a small region at low fields where the magnetoresistance is positive (not

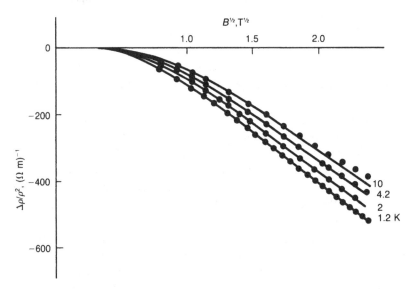

Fig. 16.3 Magnetoresistance of $Mg_{80}Cu_{20}$. The points are the experimental data and the continuous curves the theoretical fit. (After Bieri *et al.* 1986.)

Table 16.2 Data from Bieri et al. (1986)

Alloy	$Mg_{80}Cu_{20}$	$Cu_{50}Lu_{50}$	$Pd_{80}Si_{20}$	$Cu_{50}Y_{50}$	$Cu_{57}Zr_{43}$	$Y_{80}Si_{20}$
$\rho(\mu\Omega\,cm)$	52	163	78	159	180	505
$1/\tau(10^{15}\,s^{-1})$	0.80	3.33	1.78	3.00	4.33	11.22
α	1.23	1.13	1.1	1.5	1.0	1.27
$1/\tau_{in}(10^{11}\,s^{-1})$	3.74	1.64	2.27	1.17	0.46	0.52
$A(10^9\,s^{-1}\,K^{-2})$	–	–	4.1	6.9	2.5	1.07
$1/\tau_{so}(10^{12}\,s^{-1})$	0.096	16.8	11.5	1.4	1.69	2.12
$D(cm^2\,s^{-1})$	7.7	2.2	4.5	2.3	1.9	0.70

ρ is the resistivity at 4.2 K and $1/\tau$ the corresponding elastic scattering probability.
α is a 'fudge factor' inserted in front of equations (A1.4) and (A1.5) for the magnetoconductivity due to weak localisation.
$1/\tau_{in}$ is the inelastic scattering probability at 4.2 K, which varies as AT^2 at other temperatures.
$1/\tau_{so}$ is the spin–orbit scattering probability and D the diffusion coefficient, calculated on the nearly-free-electron model.

visible in the figure) implies that there is weak spin–orbit scattering and allows its lifetime to be found.

By contrast the spin–orbit scattering in $Cu_{50}Lu_{50}$ and $Pd_{80}Si_{20}$ is strong enough to keep the magnetoresistance positive at all fields in the experiments. The results are shown in Figures 16.4 and 16.5 with the corresponding theoretical comparisons. The parameters are again given in Table 16.2. For $Cu_{50}Y_{50}$ the results and comparison are shown in Figure 16.6 and the table. In this glass there are serious discrepancies at very low temperatures and high fields.

The inelastic scattering time, or more generally the phase-breaking time, is determined from the experiments, which show that above 5 K $1/\tau_{in}$ varies as T^2 and is of the right size to be attributed to phonon scattering. Below 1–2 K the variation is as $T^{1/2}$; the origin of this is not explained. The spin–orbit relaxation times for the range of alloys measured can be calculated on the assumption that the spin–orbit scattering arises from the d-states of the transition metal and not from the non-transition element (Mg,Cu,Si). This provides reasonable agreement with the values deduced from the experiments.

The overall impression from all these data is that the fit at low fields is good but tends to worsen at large values of B/T. The exception is the behaviour of the simple metal combination $Mg_{80}Cu_{20}$ which seems to give reasonable agreement over the whole range.

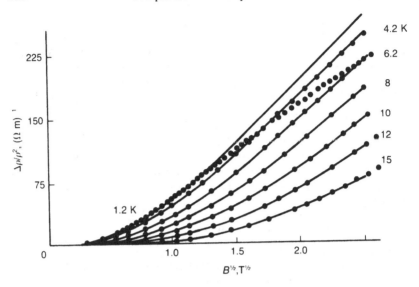

Fig. 16.4 Magnetoresistance of $Cu_{50}Lu_{50}$. The points are the experimental data and the continuous curves the theoretical fit. (After Bieri *et al.* 1986.)

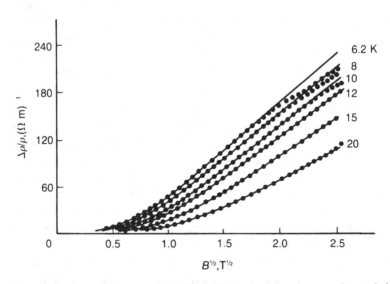

Fig. 16.5 Magnetoresistance of $Pd_{80}Si_{20}$. The points are the experimental data and the continuous curves the theoretical fit. (After Bieri *et al.* 1986.)

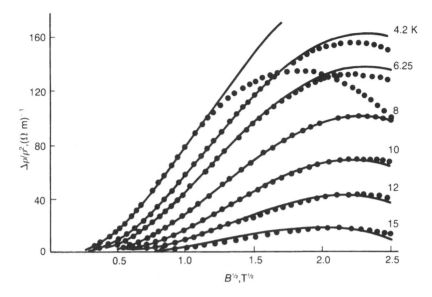

Fig. 16.6 Magnetoresistance of $Cu_{50}Y_{50}$. The points are the experimental data and the continuous curves the theoretical fit. (After Bieri *et al.* 1986.)

16.3.2 The enhanced interaction effect

The alloy $Y_{80}Si_{20}$ has by far the highest resistivity and the smallest value of the diffusion coefficient D of the glasses considered. This already suggests that the enhanced interaction effect may be important here; indeed the temperature dependence of the $B^{1/2}$ term at high fields confirms that weak localisation alone cannot account for the magnetoresistance.

The behaviour can, however, be explained as a combination of weak localisation and the enhanced interaction effect (again neglecting the effect of spin–orbit scattering), the latter increasing in importance at low temperatures and high fields. The results for this alloy are shown in Figure 16.7 in comparison with theory, calculated with $\tilde{F} = 0.62$ and the other parameters as in Table 16.2.

16.3.3 Superconducting fluctuations: Maki–Thompson correction

Bieri and colleagues also studied $Cu_{57}Zr_{43}$, which is a superconductor with a transition temperature of about $0.7\,K$. They therefore had to take account of the Maki–Thompson correction, which we referred to

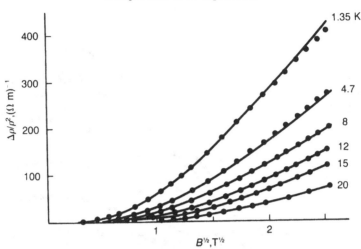

Fig. 16.7 Magnetoresistance of $Y_{80}Si_{20}$. The points are the experimental data and the continuous curves the theoretical fit. (After Bieri *et al.* 1986.)

in section 14.6 and Appendix A5. For this material it is expected that the correction $\beta(T)$ is independent of B for fields and temperatures at which $B/T \ll 0.7\,T\,K^{-1}$. The experimental results and theoretical calculations are compared in Figure 16.8; the parameters used are given in Table 16.2.

Figure 16.9 compares the values of $\beta(T)$ needed to fit the data with the curve calculated from the theory with $T_c = 0.8\,K$. At high values of B/T the correction $\beta(B, T)$ is expected to be smaller and so account for the divergence seen at high fields.

16.3.4 Magnetic impurities

By adding 0.5 atomic % of Gd to the CuY and CuLu glasses, Bieri and colleagues hoped to test the effect on weak localisation of impurities that carry a magnetic moment. The results were not what the authors expected and it has since become clear that the effect of magnetic impurities is much more subtle and complex than was realised at the time. [See note (1) of this chapter.]

16.3.5 Provisional conclusions

The outcome of these wide-ranging experiments is the conviction that the theory goes a long way to account for the experimental findings, not only qualitatively but semi-quantitatively. The use of the α-factor (see Table

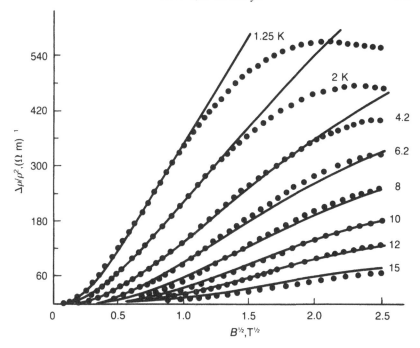

Fig. 16.8 Magnetoresistance of $Cu_{57}Zr_{43}$. The points are the experimental data and the continuous curves the theoretical fit, which includes the Maki–Thompson correction to take account of superconducting fluctuations. This is shown in Figure 16.9. (After Bieri *et al.* 1986.)

16.2) is rather unsatisfactory and the fact that the $T^{1/2}$ dependence for phase breaking cannot yet be explained raises doubts about the completeness of the theory. But this is rather to be expected at this stage of the subject (see, however, comments at the end of Appendix A.5).

The value of the wide-ranging set of experiments carried out by Bieri *et al.* (1986) lies partly in the diversity of the materials studied but also in the attempts at each stage to link the experiments with different aspects of the theory. Each aspect is then analysed to find out if the data bear a credible relationship to independently determined data, such as, for example, the spin–orbit numbers, the electron–phonon scattering rate and the $\beta(T)$ values in the Maki–Thompson correction.

16.4 Ca–Al alloys

This set of metallic glasses has proved very popular with experimentalists, partly because it can be made in a wide range of compositions, partly

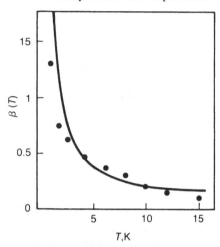

Fig. 16.9 The Maki–Thompson corrections $\beta(T)$ used in fitting the data of Figure 16.8. The continuous line is calculated from the theoretical expression of Larkin (1980).

because it provides a wide range of resistivities including some very high ones (the resistivities reach more than $400 \, \mu\Omega \, \mathrm{cm}$ at a composition of just over 40 atomic % Al) and mainly perhaps because the alloys are made from non-transition metals.

The immediate question is: why is the resistivity so high? Although Ca is next door in the periodic table to the first of the transition metals, scandium, pure Ca metal is clearly a simple metal with no occupied d-band. Nevertheless the possibility arises that alloying in the amorphous state with the trivalent metal Al might raise the Fermi level enough to reach some of the unoccupied d-states and so form a small d-band. Indeed the Fermi energy is so close to the d-states that it makes the pseudopotentials very sensitive to its details and the calculation of the resistivity of the liquid unreliable.

It is also possible that the nearness in energy of the unoccupied d-states brought about by the addition of Al may cause strong d-hybridisation of the sp-band without actually causing a separate band to form and be enough together with quantum interference to explain the big resistivity. Let us look at the experimental data with these ideas in mind.

Table 16.3 shows values of γ derived from specific-heat data, where γT is the electronic part of the low-temperature specific heat and is proportional to the density of states at the Fermi level. The measured values are rather higher than the ideal free-electron values but this is in any case to be expected from electron–phonon enhancement. The ratio of measured

Table 16.3 *Ca–Al glasses: electronic specific heat and Hall coefficient*

Alloy	$\gamma(\text{mJ mol}^{-1}\,\text{K}^{-2})$	$d(\text{g cm}^{-3})$	$R_H(10^{-11}\,\text{m}^3\,\text{C}^{-1})$	$(e/a)_{\text{eff}}$	$(e/a)_{\text{FE}}$
$\text{Ca}_{75}\text{Al}_{25}$	2.11	1.80	−26.0	0.82	2.25
$\text{Ca}_{70}\text{Al}_{30}$	1.82	1.85	−19.6	1.03	2.30
	(1.68)				
$\text{Ca}_{65}\text{Al}_{35}$	1.74	1.90	−14.4	1.34	2.35
$\text{Ca}_{60}\text{Al}_{40}$	1.67	1.96	−15.8	1.16	2.40
	(1.64)				

Data from Mitzutani and Matsuda (1983); values in parentheses from Mitzutani *et al.* (1987).
(e/a) is the electron to atom ratio; $_{\text{FE}}$ = free-electron value.

to ideal of 1.4 for $\text{Ca}_{60}\text{Al}_{40}$ is to be compared with a measured thermo-power enhancement of 1.3 (see Chapter 15). This is not unreasonable and the density of states as a function of concentration of Al shows no anom-aly. The Hall coefficient R_H is also shown in Table 16.3; it is negative and independent of temperature but its values are very different from the free-electron values. Indeed the table shows that the electron to atom ratio (e/a) deduced from R_H is only about one-half that derived from the composition. We cannot therefore assume free-electron behaviour; on the other hand, strong hybridisation with the neighbouring d-states may be able to explain the anomalous Hall coefficients without invoking an occupied d-band.

Whether or not there is a separate d-band in some or all of the alloys, it is generally thought that the number of electrons in the d-band would be small enough that they would make only a minor contribution to the conductivity.

The low-temperature behaviour of the resistivity of this alloy series has been much investigated as a function of magnetic field; another variable has been the concentration of small additions of heavy elements to increase the spin–orbit scattering. The data have been analysed in terms of weak localisation and the interaction effect in the way we have already seen and some of the recent measurements have yielded impressive agreement with the form of the theory as Figures 16.10 to 16.15 illustrate.

Sahnoune, Strom-Olsen and Fischer (1992) measured the resistivity of $\text{Ca}_{70}\text{Al}_{30}$ (about $310\,\mu\Omega\,\text{cm}$) at temperatures up to 25 K and at fields up to 10 T; they then replaced some Al with small amounts of Ag and Au to study the effects of spin–orbit scattering. The Figures 16.10–16.12 show

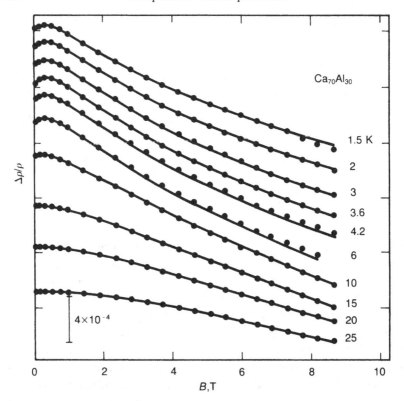

Fig. 16.10 Normalised magnetoresistance of $Ca_{70}Al_{30}$ at different temperatures. Points are from experiment; curves from the theoretical fit. (After Sahnoune *et al.* 1992.)

the field dependence of $Ca_{70}Al_{30}$ and the ternary alloys containing two different levels of Au as examples.

Mayeya and Howson (1992) chose to study $Ca_{80}Al_{20}$ because of its comparatively low resistivity (about 140 μΩ cm and k_Fl about 2.5) so that they could try to satisfy the condition $k_Fl \gg 1$ and still clearly see the effects of weak localisation. They also wished to contrast this alloy with the high-resistance alloy $Ca_{60}Al_{40}$ (about 330 μΩ cm). They added small concentrations of Ag and Au to the two host alloys to bring into prominence the effects of spin–orbit scattering. Figures 16.13–16.15 show their data on $Ca_{80}Al_{20}$ by itself and also alloyed with 0.35 and 1.34 atomic % Au.

In these two sets of experiments, three host and seventeen dilute ternary alloys were measured with concentrations of up to nearly 6 atomic %

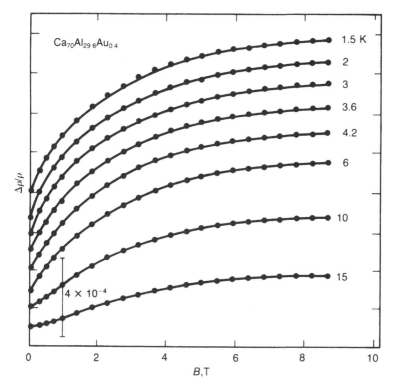

Fig. 16.11 Normalised magnetoresistance of $Ca_{70}Al_{29.6}Au_{0.4}$ at different temperatures. Points are from experiment; curves from the theoretical fit. (After Sahnoune *et al.* 1992.)

Au and 2 atomic % Ag. As the data in Table 16.4 show, the change in resistivity with the addition of Ag or Au is not monotonic and the reason for this is not clear. In the absence of heat-capacity data to give us the density of states at the Fermi level, this makes the derivation of the appropriate values of D uncertain, as we shall see below.

All the results for the magnetic field dependence were found to conform to the curve derived from theory as closely as the examples shown. From the analysis the dephasing time τ_{in}, the spin–orbit scattering time τ_{so} and the interaction parameter F were determined. The values are collected in Table 16.4; to test the theory we need to satisfy ourselves that not only do the functional forms fit the data but the parameters are reproducible and make physical sense.

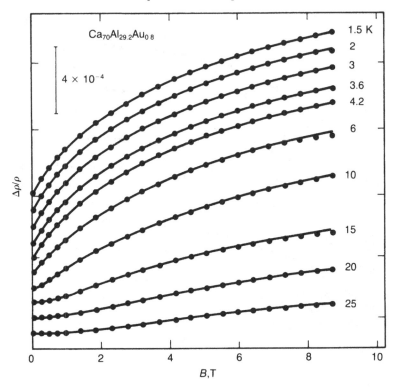

Fig. 16.12 Normalised magnetoresistance of $Ca_{70}Al_{29.2}Au_{0.8}$ at different temperatures. Points are from experiment; curves from the theoretical fit. (After Sahnoune 1992.)

16.4.1 Values of D

The diffusion coefficient D is found from the Einstein relation, which requires a knowledge of the density of states. This has been measured only for the host alloys and some assumption must then be made to find D for the dilute ternary samples. In one case the authors assumed that the density of states remained unchanged and in the other that D remained unchanged on alloying. The value of D does not influence the quality of fit between experiment and theory as shown in the figures but does alter the absolute magnitude of the scattering times. To illustrate this I have calculated D in the data from source (1) in Table 16.4 on the basis of unchanged density of states and given the values in brackets in Table 16.4. I also include the corresponding values of τ_{so}. On the other hand, the value of D used by Sahnoune *et al.* (1992) suggests that they were

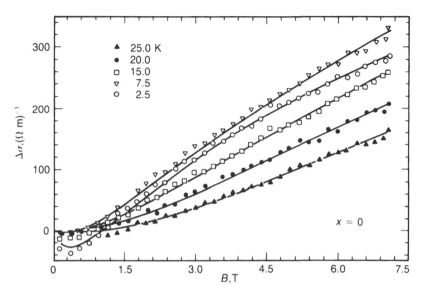

Fig. 16.13 Magnetoconductivity of $Ca_{80}Al_{20}$ at different temperatures. Points are from experiment; curves from the theoretical fit. (After Mayeya and Howson 1992.)

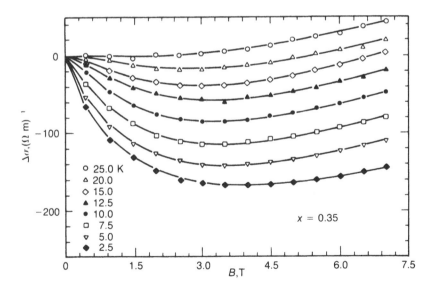

Fig. 16.14 Magnetoconductivity of $Ca_{80}Al_{20}$ with 0.35 atomic % Au added. The different curves correspond to different temperatures. Points are from experiment; curves from the theoretical fit. (After Mayeya and Howson 1992.)

Table 16.4 Characteristic parameters of Ca–Al–(Ag,Au) amorphous alloys

Alloy	$\rho(\mu\Omega\,cm)$	$D(10^{-5}\,m^2\,s^{-1})^b$	$F(B)$	$F(T)$	τ_{so} (ps)	Source[a]
$(Ca_{80}Al_{20})_{100-x}M_x$						
$x = 0$	167	9.6 (9.6)^b	0.25		33	
M = Ag						
$x = 0.5$	130	9.6 (12)	0.23		22 (18)	
$x = 2.5$	122	9.6 (13)	0.10		15 (11)	
M = Au						(1)
$x = 0.35$	133	9.6 (12)	0.10		5.8 (4.6)	
$x = 1.3$	159	9.6 (10)	0		1.7 (1.6)	
$x = 2.9$	133	9.6 (13)	0		1.0 (0.7)	
$x = 5.9$	168	9.6 (9.5)	0		0.9 (0.9)	
$(Ca_{60}Al_{40})_{100-x}Au_x$						
$x = 0$	330	6.7	0.10		73	
$x = 0.2$	271	6.7 (8.2)	0.05		8.2 (6.7)	
$x = 0.5$	310	6.7 (7.1)	0.05		1.9 (1.8)	

Ca$_{70}$Al$_{30-x}$M$_x$						
$x = 0$	310	15 (10)	0.56	0.18	12.2 (18)	
M = Ag						
$x = 0.3$	264	18 (12)	0.47	0.22	6.1 (9)	
$x = 0.7$	270	17 (11)	0.40	0.19	4.3 (6.5)	
$x = 2$	245	19 (13)	0.24	0.08	1.8 (2.7)	(2)
M = Au						
$x = 0.1$	297	16 (11)	0.19	0.02	1.6 (2.4)	
$x = 0.2$	280	17 (11)	0.10	0.01	0.76 (1.1)	
$x = 0.4$	280	17 (11)	0.08	−0.08	0.41 (0.6)	
$x = 0.8$	290	16 (11)	0.02	−0.08	0.21 (0.3)	
$x = 2$	220	21 (14)	0.0	−0.11	0.14 (0.2)	
$x = 3$	210	23 (15)	0.0	−0.25	0.07 (0.1)	
$x = 0$	308	15 (10)	0.29		8.6	(3)
Ca$_{75}$Al$_{25}$						
$x = 0$	261	9.1			57	
Ca$_{65}$Al$_{35}$						
$x = 0$	582	4.5			110	(4)

[a] (1) Mayeya and Howson (1992); (2) Sahnoune et al. (1992), where the values of F given are in fact $\tilde{F}$, they can be changed into F approximately by adding $\tilde{F}/12$. (3) Lindqvist et al. (1990); (4) Gey and Weyhe (1992).

[b] Values of D in brackets relating to data from source (1) are derived from the enhanced density of states on the assumption that $N(E_F)$ is *unchanged on alloying*. Those relating to sources (2), (3) are derived on the same assumption from the *enhanced* density of states, not the bare value used by the authors. Bracketed values are thus derived on a consistence basis and show broad agreement.

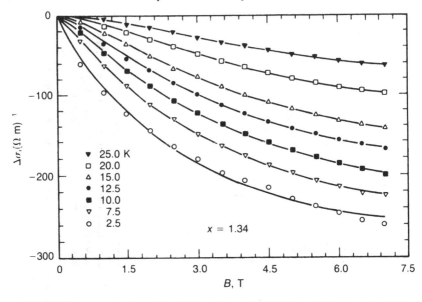

Fig. 16.15 Magnetoconductivity of $Ca_{80}Al_{20}$ with 1.34 atomic % Au added. The different curves correspond to different temperatures. Points are from experiment; curves from the theoretical fit. (After Mayeya and Howson 1992.)

using the bare density of states and that their values of D are a factor of about 1.5 too high. If their values of τ are corrected (i.e. increased by a factor of 3/2 as shown in the bracketed values) there is much more consistency between the different sets of data in the table.

16.4.2 Spin–orbit scattering

The spin–orbit scattering rate was found to increase with concentration of heavy element (Ag or Au). At lower concentrations the spin–orbit scattering rate is proportional to the concentration of the heavy element and the ratio of the slopes for Au and for Ag is about 20 in both sets of experiments. Mayeya and Howson (1992) point to a possible explanation of this ratio. A calculation of the matrix element for spin–orbit coupling in an atom, using hydrogen-like wavefunctions, shows that it should vary as Z^4/n^3 where Z is the atomic number (nuclear charge) and n is the principal quantum number of the orbital involved. Thus the scattering rates here should scale as Z^8/n^6 with $Z = 79$ and $n = 5$ for Au; for Ag, $Z = 47$ and $n = 4$, so that the ratio is about 20, in excellent agreement with the experiments.

16.4.3 The value of F

Because it is possible to isolate the consequences of the enhanced inter-action effect fairly accurately from those due to weak localisation, the experiments by Sahnoune *et al.* (1992) on $Ca_{70}Al_{30}$ are able to demon-strate how spin–orbit scattering alters the magnetoresistance due to enhanced electron interaction. Sahnoune *et al.* interpret this as reducing the effective value of $\tilde{F}$, as shown in the table and also in Figure 16.16. This was probably done more for convenience than a belief that this represented the physics of what was happening.

What is clear, however, before any interpretation is made, is the very important point that adding small concentrations of Ag or Au can elim-inate the magnetoresistance associated with the interaction effect within the field range explored.

According to our discussion in section 14.3, we would interpret the result as meaning that spin–orbit scattering disrupts the interference of the electron pairs with antiparallel spins and that the effects are to leave $\tilde{F}$ unchanged (to a first approximation) but to alter the effective tempera-ture in the expression for the magnetoresistance. If we use the values of τ_{so} derived from weak localisation, this interpretation appears to work as far as it is possible to judge from the data in the published diagrams. One would also expect the corresponding temperature-dependent contribu-tion to be altered and indeed this is found (see below).

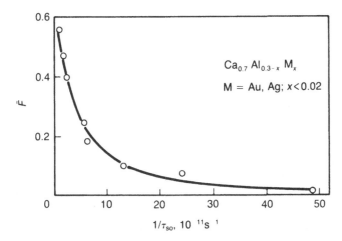

Fig. 16.16 $\tilde{F}$ as a function of $1/\tau_{so}$. (Data from Sahnoune *et al.* 1992 as replotted by Rapp 1993.)

16.4.4 Dephasing times τ_{in}

The temperature dependence of $1/\tau_{in}$ was determined from the effects of weak localisation for both $Ca_{80}Al_{20}$ and $Ca_{70}Al_{30}$ and their ternary counterparts. The important feature is that for a given host all the specimens give very similar results almost independent of the added Ag or Au; this is to be expected on the grounds that the host material is not likely to be much changed by small additions of impurity. The results thus demonstrate the consistency of the measurements.

The $Ca_{70}Al_{30}$-based group of alloys show that at higher temperatures $1/\tau_{in}$ varies as T^3 while the $Ca_{80}Al_{20}$-based group give a T^2 dependence. Both are only approximate power laws and the difference is not as pronounced as that bald statement implies. Both can be and have been attributed to phonon scattering.

At low temperatures τ_{in} becomes independent of temperature. This is attributed to spin–flip scattering from residual magnetic impurities and comes about as follows. The inelastic scattering by phonons, which destroys the enhanced resistance due to weak localisation, falls off as the temperature is reduced until the associated coherence time τ_{in} becomes comparable with the spin–flip scattering time τ_s. Then spin flips, which alter the spin direction of the scattered electron, begin to dephase the coherent wavefunctions, whether with parallel or antiparallel spins. Thus the dephasing time becomes independent of temperature and assumes the value τ_s. The value determined from the experiments appears to be consistent with the concentration of residual magnetic impurities in the host materials (at the level of parts per million).

16.4.5 Analysis of the temperature dependence

The final part of the analysis concerns the temperature dependence of the resistivity, which so far has been carried out only on their data by Sahnoune *et al.* (1992). The low-temperature end of this, where we expect the enhanced electron interaction effect to dominate, is illustrated in Figure 16.17, where $\Delta\rho D^{1/2}/\rho^2$ is plotted against $T^{1/2}$. Though the data fit this power law approximately, the values of $\tilde{F}$ so determined are very different from those found from the field dependence and in the high-concentration alloys have to be negative. This discrepancy of an inconsistency between the values of F found from field dependence and temperature dependence has been found in other systems.

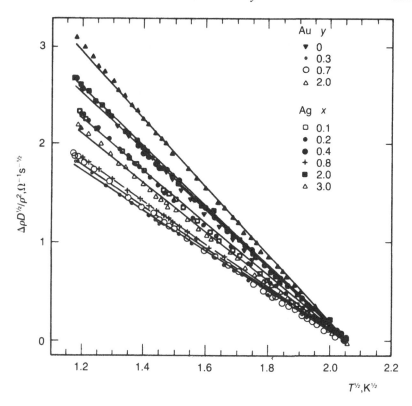

Fig. 16.17 Temperature dependence of $\Delta\sigma$ for $Ca_{70}Al_{30}$ and alloys with Ag and Au. (After Sahnoune *et al.* 1992.)

The analysis shows clearly that the temperature dependence of the ternary alloys, in this region where the interaction effect is dominant, is altered by the addition of a heavy element, which implies that the change is due to spin–orbit scattering. The argument given in section 13.3 suggests that the excess resistance due to the enhanced interaction effect does not vary exactly as $T^{1/2}$ unless τ_{so} is much less than or much greater than τ_T. The coefficients of $T^{1/2}$ are different in the two cases. Qualitatively this causes a positive curvature to the plot of $\Delta\sigma_{int}$ versus $T^{1/2}$. This is indeed seen in Figure 16.17 although it is not possible to make a reliable quantitative comparison. Nonetheless there is no doubt that spin–orbit scattering alters the contribution to the conductivity of the enhanced electron interaction and, at least qualitatively, it behaves as one would expect.

16.5 Conclusions

We have looked at a very limited number of the experimental data available; there have been very many measurements to show that the theory of weak localisation and the enhanced interaction effect can explain the main features of the electron transport in metallic glasses. There is no doubt that the unusual temperature dependence of the conductivity of metallic glasses, for example its increase with increasing temperature and its alteration by the addition of small concentrations of heavy elements, the size, isotropy and field dependence of the magnetoresistance all conform to the predictions of theory. The theory applies equally to simple, non-transition metal alloys as to those containing transition metals but the former are at present more amenable to theoretical comparison because the carriers are confined to a single band and thus their numbers are known. The extension and modification of Boltzmann theory to meet the circumstance of strong elastic scattering have been very successful. The introduction of the simple notion of diffusion by conduction electrons and the exploitation of the classical theory of diffusion together with the subtleties of the interference of electron waves have opened up the whole study of disordered metals which had previously seemed so opaque and intractable.

The final question of whether the theory is fully complete is still to be answered. It will certainly continue to develop under the impulse of more exact and more exacting measurements.

Appendices

A1 Some formulae in weak localisation

A1.1 Temperature dependence

The very striking effects of spin–orbit scattering are summed up in the following expression for the conductivity due to weak localisation in the presence of spin–orbit scattering:

$$\Delta\sigma(T) \simeq (e^2/4\pi^2)D^{-1/2}[3\{1/\tau_{in} + 4/\tau_{so}\}^{1/2} - 3(4/\tau_{so})^{1/2} - (1/\tau_{in})^{1/2}] \quad (A1.1)$$

where τ_{in} is the dephasing, often the inelastic, scattering time. If $\tau_{so} \gg \tau_{in}$ so that there is effectively no spin–orbit scattering, $\Delta\sigma$ is positive and the square bracket reduces to $2/(\tau_{in})^{1/2}$. On the other hand, if τ_{so} becomes short, the term in $1/\tau_{so}$ dominates the expression in curly brackets which becomes constant. The temperature dependence is thus dominated by the term $-(1/\tau_{in})^{1/2}$, which gives a negative temperature coefficient of conductivity, thus reversing the previous state of affairs.

A1.2 Magnetic field dependence

The expression for the change in conductivity at low magnetic fields due to weak localisation, in the absence of spin–orbit effects (see below), is:

$$\Delta\sigma(B) = (e^2/12\pi^2\hbar)(D\tau_{in})^{3/2}(eB/\hbar)^2 \quad (A1.2)$$

and at high fields:

$$\Delta\sigma(B) = (0.605e^2/2\pi^2\hbar)(eB/\hbar)^{1/2} \quad (A1.3)$$

I have written these results in terms of the change in conductivity rather than as relative magnetoresistance partly because the formulae are perhaps simpler in this form but also because that is the way they are usually presented in the literature. Experimentalists usually measure $\Delta R/R = \Delta\rho/\rho$ and so you often see $\Delta\rho/\rho^2$, which is just $-\Delta\sigma$, plotted against magnetic field or quoted in formulae in experimental papers.

The expression (A1.2) is not physically realistic since there is always some spin–orbit scattering and this must manifest itself at low enough fields. The more general expressions that take account of spin–orbit scattering are as follows:

225

$$\Delta\sigma(B) = -(e^2/4\pi^2 h)(eB/h)^{1/2}[f(\hbar/4eDB\tau_{in}) - 3f(\hbar/4eDB\tau_{in}^*)] \qquad (A1.4)$$

where $1/\tau_{in}^* = (1/\tau_{in} + 2/\tau_{so})$ and the function f has the following limiting behaviour:

$$f(x) = 0.605 \text{ for } x \ll 1; \ f(x) = x^{3/2} \text{ for } x \gg 1.$$

The sign of $\Delta\sigma$ depends, as we have seen, on the relationship of the magnetic field lifetime to the spin–orbit scattering time and the dephasing time τ_{in}.
We distinguish four regimes:

1 At low magnetic fields the conductivity, as we have seen, varies as B^2. If there were no spin–orbit scattering the magnetoconductivity would be positive, giving the unusual negative magnetoresistance as discussed in section 11.7.1.

$$\Delta\sigma(B) = +(e^2/96\pi^2 \hbar)(4D\tau_{in})^{3/2}(eB/\hbar)^2 \qquad (A1.5a)$$

2 If, however, there is spin–orbit scattering, there is necessarily a low-field regime where $\tau_B > \tau_{so}$ and so the magnetoresistance is positive; then we have:

$$\Delta\sigma(B) = -(e^2/192\pi^2 \hbar)(4D\tau_{in})^{3/2}(eB/\hbar)^2 \qquad (A1.5b)$$

3 At higher fields we can move into the $B^{1/2}$ regime but, if we still have $\tau_B > \tau_{so}$, we continue with the positive magnetoresistance (negative magnetoconductivity):

$$\Delta\sigma(B) = -(0.605e^2/4\pi^2 \hbar)(eB/\hbar)^{1/2} \qquad (A1.5c)$$

4 Finally at very high fields $\tau_B < \tau_{so}$ and so the spin–orbit scattering does not have time to operate before the field has dephased the wavefunctions. Then spin–orbit scattering is ineffective and the magnetoresistance becomes negative again:

$$\Delta\sigma(B) = +(0.605e^2/4\pi^2 \hbar)(eB/\hbar)^{1/2} \qquad (A1.5d)$$

Equation (A1.4) derives originally from Altshuler *et al.* (1981) and is widely used. It has been discussed by Lindqvist and Rapp (1988), who point out that it differs significantly from the results of Fukuyama and Hoshino (1981). This difference arises because the latter put in the so-called Zeeman contribution, discussed in section 14.5. The difference disappears when D is small; it also disappears when D is greater than $10^{-4} \text{ m}^2 \text{ s}^{-1}$ but there are regions of practical importance where the Zeeman contribution is significant.

A2 Weak localisation with s- and d-bands

We have seen in our discussion of alloys that contain transition metals that there are two groups of conduction electrons, the s- and d-electrons, the former almost certainly strongly hybridised with p- and d-character. We must now ask: How does this affect the weak localisation and mechanism we have been discussing?
    The answer is that the formalism developed for a single isotropic band is valid for an anisotropic, multi-band alloy provided that the characteristic parameters, D the diffusion coefficient, $N(E_F)$ the density of states and the relaxation times are all properly interpreted. (See Rainer and Bergmann 1985.)

We denote the density of states N, conductivity σ, elastic scattering time τ and diffusion coefficient D of the two bands by the subscripts s and d. The probability of finding an electron in band s is then:

$$p_s = N_s/(N_s + N_d) \qquad (A2.1)$$

and correspondingly for p_d. In a time interval t, an electron thus spends a time $t_s = p_s t$ in the s-band and $p_d t$ in the d-band. In the time t_s the electron diffuses a distance (in three dimensions):

$$(r_s)^2 = 3D_s t_s \qquad (A2.2)$$

Correspondingly for band d. Since in diffusive motion the square of the distances add, the total distance travelled is:

$$r^2 = (r_s)^2 + (r_d)^2 = 3(p_s D_s + p_d D_d)t = 3Dt$$

$$\text{where } D = (N_s D_s + N_d D_d)/(N_s + N_d) \qquad (A2.3)$$

The total conductivity is given by:

$$\sigma = \sigma_s + \sigma_d = e^2(D_s N_s + D_d N_d) = e^2 D N \qquad (A2.4)$$

$$\text{where } N = N_s + N_d.$$

If the inelastic scattering times are different in the two bands, the effective probability of inelastic scattering is an average weighted according to the time spent in each band:

$$1/\tau_{in} = p_s(1/\tau_{in,s}) + p_d(1/\tau_{in,d}) \qquad (A2.5)$$

Similar results apply to the spin–orbit and other scattering times. In all this we are assuming that the electrons change bands often enough to sample their different properties impartially.

The upshot is that the formalism of weak localisation is valid provided that the parameters are interpreted according to equations (A2.3), (A2.4) and (A2.5) with appropriate generalisations of these in more complex systems.

There is no reason why the properties of the two bands should not differ widely. For example, if the s-electrons were purely s-like, they would undergo no spin–orbit coupling in the s-band and only when scattered into the d-band would they suffer spin–orbit scattering. This extreme situation does not arise in practice because the so-called s- and d-wavefunctions are hybrids, but there could still be a marked difference in spin–orbit scattering in the two bands. This means that changing the concentration of the components can alter the spin–orbit scattering not only because the relative number and strength of spin–orbit scatterers change but also because the proportions of s- and d-like electrons can alter.

A3   Fourier transform of the classical diffusion probability

We wish to find the Fourier transform of the probability:

$$P(r, t) = [\exp(-r^2/4\pi Dt)]/(4\pi Dt)^{3/2} \qquad (A3.1)$$

To do so we proceed somewhat indirectly and use the equation of continuity for n particles per unit volume at position $\mathbf{r}$ and time t, forming a current of particles of density $\mathbf{j}$:

$$\partial n/\partial t = S(\mathbf{r}, t) - \text{div} \mathbf{j} \qquad (A3.2)$$

This says that the rate of change of n in an infinitesimal volume equals the rate at which particles are produced (the source term S) minus the rate they flow out (the div term). If the particles move by diffusion:

$$\mathbf{j} = -D \text{ grad } n \qquad (A3.3)$$

and
$$\text{div } \mathbf{j} = -D\nabla^2 n \qquad (A3.4)$$

If the source produces a particle at $t = t_0$ and $\mathbf{r} = \mathbf{r}_0$, S becomes the product of two δ-functions: $S = \delta(\mathbf{r} - \mathbf{r}_0)\delta(t - t_0)$, which represents the effect of producing a single particle repeatedly and then averaging. Thus we have for equation (A3.2):

$$\partial n/\partial t - D\nabla^2 n = \delta(\mathbf{r} - \mathbf{r}_0)\delta(t - t_0) \qquad (A3.5)$$

We now take the Fourier transform of this equation, turning $n(\mathbf{r}, t)$ into $n(\mathbf{q}, \omega)$; the Fourier transform of the δ-functions is just unity while that of the left-hand side of equation (A3.5) is $(-i\omega + Dq^2)n(\mathbf{q}, \omega)$.

We therefore get:

$$(-i\omega + Dq^2)n(q, \omega) = 1 \qquad (A3.6)$$

so that finally we have:

$$n(q, \omega) = 1/(Dq^2 - i\omega) \qquad (A3.7)$$

or
$$n(q, \omega) = (i\omega + Dq^2)/[\omega^2 + (Dq^2)^2] \qquad (A3.8)$$

This gives us the classical probability per unit frequency range of finding a particle with Fourier components q and ω when the particle is free to diffuse with a diffusion coefficient D.

A4 Modified values of F

A4.1 Density of states

It is easy to see that F is often not small compared to unity. In a typical alloy with one or two conduction electrons per ion, k_F is typically of order $1/a$ where a is the ionic separation. The reciprocal of the screening radius χ is also roughly $1/a$ and so x in equation (12.37), which is their ratio, is about 1. This makes $F = \ln 2$, which is not small compared to unity.

When F is not small compared to unity, a more exact treatment of screening at small values of q is needed to take account of diffusive effects when the two interacting electrons are far apart. The form of the density of states is unchanged and only the coupling strength is altered. Moreover, this new coupling parameter F^* can still be expressed in terms of F and becomes:

$$F^* = 4[(1 + F/2)^{1/2} - 1] \qquad (A4.1)$$

As F tends to unity, this tends to the original value F. When $F = 1$, its maximum value, $F^* = 0.9$ so the differences introduced are surprisingly small.

The total coupling strength with the correction is:

$$g_{\text{total}} = 2\{1 - 3[(1 + F/2)^{1/2} - 1]\} \qquad (A4.2)$$

A4.2 Electrical conductivity

A4.2.1 The total change in conductivity in the particle–hole channel

A full calculation of the contributions to the change in conductivity from both exchange and Hartree terms in the particle–hole channel gives:

$$\Delta\sigma(T) = 0.915(e^2/4\pi^2\hbar)(4/3 - 3\widetilde{F}/2)(k_B T/\hbar D)^{1/2} \tag{A4.3}$$

The coupling strength in the Hartree processes is not exactly F but $\widetilde{F}$ given by:

$$\widetilde{F} = 32[(1 + F/2)^{3/2} - 1 - 3F/4]/3F \tag{A4.4}$$

The rather complex form of $\widetilde{F}$ arises as before because the interaction between the electron pairs is altered by the diffusive motion of the screening electrons. Since the correction to F is always small, an expansion of the terms in the round bracket of equation (A4.4) shows that to an approximation adequate for many purposes:

$$\widetilde{F} = F(1 - F/12) \tag{A4.5}$$

For example, when $F \to 0$, $\widetilde{F} \to F$; when $F = 1$, $\widetilde{F} = 0.93$ according to the exact formula and $\widetilde{F} = 0.92$ according to the approximation.

A4.3 Magnetoresistance due to the enhanced interaction effect

In quoting the final results it is convenient to give here only those for the spin-splitting or Zeeman effect; results including the orbital contribution can be found, for example, in Lindqvist and Rapp (1988).

The change in conductivity due to spin-splitting at high fields $(g\mu B \gg k_B T)$ is:

$$\Delta\sigma(B) = (e^2/4\pi^2\hbar)\widetilde{F}(k_B T/2\hbar D)^{1/2}[(g\mu B/k_B T)^{1/2} - 1.294]$$
$$\simeq (e^2/4\pi^2\hbar)\widetilde{F}(g\mu B/2\hbar D)^{1/2} \tag{A4.6}$$

At low fields we have:

$$\Delta\sigma(B) = (e^2/4\pi^2\hbar)\widetilde{F}(0.056g\mu B/k_B T)^2 \tag{A4.7}$$

where $\widetilde{F}$ is defined in equation (A4.4).

A5 The Maki–Thompson correction

So far we have assumed that all contributions from the enhanced electron interaction that involve interference are killed off by the energy difference (typically $k_B T$) between the two interacting electrons; this leads to the characteristic $T^{1/2}$ temperature dependence. In some circumstances, however, other dephasing mechanisms can be important. This is true of the Maki–Thompson correction (Maki 1968, Thompson 1970), which is a contribution from the Cooper channel in addition to that which alters the density of states.

As we have seen, the interactions in the Cooper channel come from coherence between pairs of counter-propagating electrons executing the same closed path. For this reason, the change in conductivity that it causes resembles that due to weak localisation, except that: (1) The two electrons need a coupling mechanism; for the partial waves in weak localisation this is intrinsic. (2) Because the coupling is due to the superconducting mechanism, only the antiparallel spin pairs (in the singlet state) participate and the coupling depends on temperature.

The coupling, usually designated $\beta(T)$, depends on temperature through the parameter $\lambda_c = (\ln T_c/T)^{-1}$ (T_c is the superconducting transition temperature), a parameter we have already met. Although in three dimensions the correction to the conductivity in zero field is not seen, it provides a potentially important correction to the magnetoresistance. If the applied field B is small we expect the correction $\beta(T)$ to depend only on temperature and not on B but, since the field tends to favour parallel spin combinations, it can at higher fields diminish the strength of the coupling, which, as we saw, depends on antiparallel spin combinations.

The function $\beta(T)$ has the following limiting forms (for the low fields just discussed):

$$\text{When } |\ln T_c/T| \gg 1, \quad \beta(T) = \pi^2/6\ln^2 T_c/T \quad \text{(A5.1)}$$

$$\text{When } -\ln T_c/T \ll 1, \quad \beta(T) = -\pi^2/4\ln T_c/T \quad \text{(A5.2)}$$

In between these limits, $\beta(T)$ has been tabulated (Larkin 1980); it is shown for a particular alloy as a function of T in Figure 16.9. To give some idea of its size in general, we note that when $T = 2.7T_c$, $\beta(T) \simeq 1$. It increases rapidly as T approaches T_c. From our discussion of the effect of spin–orbit scattering on weak localisation, it is clear that, unlike the triplet state, the singlet state, and hence the Maki–Thompson correction, are unchanged by such scattering. Finally the magnetoconductivity due to weak localisation in the presence of spin–orbit scattering and including the Maki–Thompson correction is:

$$\Delta\sigma(B) = (e^2/2\pi^2\hbar)(eB/\hbar)^{1/2}\{3f(4DeB\tau_{in}^*/\hbar) - [1 + 2\beta(T)]f(4DeB\tau_{in}/\hbar)\} \quad \text{(A5.3)}$$

where τ_{in} is the phase breaking or inelastic scattering time and

$$1/\tau_{in}^* = 1/\tau_{in} + 2/\tau_{so}$$

(The term in $1/\tau_{so}$ has a different coefficient for different definitions of τ_{so}.) The function f, which comes from the theory of weak localisation, is defined in Appendix A1.

Notice that only the second term in the curly brackets, which is the singlet term in weak localisation, is affected by the Maki–Thompson correction. In this term the correction simply adds to that from weak localisation.

At higher fields, strong enough to suppress spin–orbit effects, we get:

$$\Delta\sigma(B) = (0.605e^2/2\pi^2\hbar)(eB/\hbar)^{1/2}[1 - \beta(T)] \quad \text{(A5.4)}$$

which is valid for fields in the $B^{1/2}$ realm.

This discussion suggests that there could also be a correction to the magnetoresistance arising from the Hartree terms in the particle–hole channel. This would come from scattering that causes incoherence in the electronic wavefunction in addition to thermal incoherence.

Such a correction would, however, be quite different from that just discussed because in these Hartree terms, the magnetoresistance arises from Zeeman splitting rather than the orbital effect. Moreover the coupling constant is F and so does not vary with temperature or field but the correction would be modified by spin–orbit scattering. In general, however, we assume that thermal incoherence is dominant (τ_T about 10^{-12} s at 1 K) and that other sources of incoherence can here be ignored.

A6   Electron–electron scattering in disordered metals

There is another effect of electron–electron interaction in a disordered metal that is important, not for the interaction effect itself, but because it has possible consequences for weak localisation. This is the scattering of electrons by each other; even if this causes no significant resistance, it can cause dephasing and so alter the temperature dependence of the resistance. (See, for example, Kaveh and Wiser 1984.)

In section 7.1 we saw that electron–electron collisions varied with temperature as T^2 in ordered metallic materials. We are now in a position to see how this temperature dependence is altered in a disordered metal. We have already seen that the self-energy S_E of interacting electrons is complex and that the imaginary part gives rise to a finite lifetime of the interacting state. So we can use the imaginary part of the self-energy to find the lifetime of an electron interacting with another and thus get the probability of electron–electron scattering. The imaginary part of the self-energy is given in equation (A3.8). From this we deduce that the probability of electron–electron scattering through the Coulomb interaction is given by:

$$h/\tau_{ee} = \operatorname{Im} S_E$$

$$= -[\pi h N(0)(2\pi)^3]^{-1} \int N(\omega)\mathrm{d}\omega \int V(0,0)[\omega/\{(Dq^2)^2 + \omega\}]4\pi q^2 \mathrm{d}q \qquad \text{(A6.1)}$$

In dealing with the density of states we were dealing with the particle–hole interaction whereas here we wish to study the interaction of two particles with energies above the Fermi energy so the limits of integration are different. At absolute zero, one electron is given energy ϵ, the rest being below the Fermi level. This electron interacts with another electron and excites it above E_0 to an energy ϵ', which must therefore lie between 0 and ϵ. Consequently $\hbar\omega$, the difference in energy between the two, must also lie between 0 and ϵ. As before we change the variable to $y = (D/\omega)^{1/2}q$ with the range of q limited to $(\omega/D)^{1/2}$; the integral over q now becomes a definite integral with a factor $\omega^{1/2}/D^{3/2}$ outside:

$$h/\tau_{ee} \propto (D)^{-3/2} V(0,0) h^{-1} \int_0^{\epsilon/\hbar} \omega^{1/2} \mathrm{d}\omega \qquad \text{(A6.2)}$$

and finally:

$$h/\tau_{ee} \propto (\epsilon/k_F l)^{3/2}/(E_F)^{1/2} \qquad \text{(A6.3)}$$

Thus the disorder changes the energy dependence of the scattering probability from ϵ^2 to $\epsilon^{3/2}$ and the corresponding temperature dependence of electron–electron scattering from T^2 to $T^{3/2}$. In the regime where these results are valid ($\epsilon \ll E_0$), the 3/2 power law implies a stronger interaction than that of the square law. Nonetheless it is still small in the sense that h/τ_{ee} is much less than the energy of the electron so that the quasi-particle concept retains its validity.

The total scattering rate due to electron–electron collisions in a disordered alloy contains both these power laws: the T^2 term arising from the large-angle scattering and the $T^{3/2}$ term from the scattering at small values of q.

Finally therefore we can write:

$$1/\tau_{ee} = A'T^2 + B'T^{3/2} \tag{A6.4}$$

where A' and B' are positive constants.

Notes

Chapter 1

1. A detailed account of the historical development of the electron theory of metals is given in chapters 2 and 3 of *Out of the Crystal Maze – Chapters from the History of Solid State Physics* edited by Hoddeson, Braun, Teichmann and Weart and published by Oxford University Press in 1992. The history includes references and useful summaries of the physics involved.
2. There are many books on the electrical properties of metals and alloys, both in specialist accounts and as part of the broader context of solid state physics. Here are a few that deal only with metals and are not too advanced; they are concerned with Boltzmann-type theories: Cottrell (1988); Dugdale (1977); Mott and Jones (1936).
   A book that deals in detail with the electrical resistance of concentrated alloys, which may also be inhomogeneous, is by Rossiter (1987).

Chapter 2

1. A history of the methods of production of metallic glasses is given by P. Duwez (1981). In the same publication *Glassy Metals 1* (1981) and its successor *Glassy Metals 2* (1983) by the same editors and publishers there are other articles of interest though some are, not surprisingly, out of date. For a general account of the properties of metallic glasses, see, for example, the article by Cahn (1980). For accounts of disordered materials in general, see, for example, Elliot (1984) and Ziman (1979).

Chapter 3

1. See note 2 in chapter 1.

Chapter 4

1. An excellent account of the pseudopotential and of screening is given by Cottrell (1988).

234 *Notes*

Chapter 5

1. A useful review of the subject of this chapter is given by Faber (1969).

Chapter 9

1. See the review by Movaghar and Cochrane (1991).
2. The s–d hybridisation theory has been largely developed by Morgan and coworkers. See, for example, Nguyen-Mahn *et al.* (1987), which contains references to important earlier papers.

Chapter 11

1. I found papers on weak localisation in thin films by Bergmann (1983, 1984) very helpful and clear. A review [entitled *Localisation – Theory and Experiment*] has recently been presented by Kramer and MacKinnon (1993). See also the list of reviews in note 1 of Chapter 12, which, although concerned primarily with interacting electrons, also deal with weak localisation.
2. This explanation which ascribes weak localisation to the possibility of closed electron paths, with the electron wavefunction thus able to execute the same path in opposite senses and so double the classical probability, was given by Khmelnitzkii (1984) some considerable time after the original discovery had been made by diagrammatic techniques; Khmelnitzkii's paper gives an account of how this came about.
3. For a general non-mathematical account of localisation and interaction in the context of scaling theories and the metal–insulator transition, see Altshuler and Lee (1988).

Chapter 12

1. There are a number of reviews of the enhanced electron interaction (often including weak localisation). Those that concentrate on theory include: Altshuler and Aronov (1985); Fukuyama (1985); Lee and Ramakrishnan (1985). One that is more concerned with experimental data is by Howson and Gallagher (1988). A review that is concerned with the transition from metal to insulator in non-crystalline systems but which contains material pertinent to the subject of this book is by Mott and Kaveh (1985).
2. We assume that the interaction depends only on the energy difference between the states involved because for free-electron states only time *intervals* matter, not the zero of time. This means that only energy differences are significant, not the origin of energy. In the perturbed states, however, this is no longer true, as we shall see: the origin of energy is then the Fermi level.
3. The review of electron–electron scattering in conducting materials by Kaveh and Wiser (1984) provides a very clear account of many aspects of electron–electron interaction that are relevant to this book. I found their treatment of

the exchange contribution to the density of states due to the enhanced interaction effect most helpful and have followed their path.
4. This account makes no attempt to do justice to the full treatment of the Hartree contribution in the particle–hole channel. The reader is referred to the paper by Bergmann (1987).

Chapter 13

1. See the treatment by Bergmann (1987).

Chapter 14

1. The treatment of the enhanced interaction effect given by Fukuyama (1985) differs in many respects from that of Altshuler and Aronov (1985), particularly in the treatment of the coupling constants.

Chapter 15

1. A brief and clear introduction to the calculation of thermopower enhancement is given by Kaiser (1982).

Chapter 16

1. Work on the magnetoresistance of metallic glasses containing magnetic impurities has recently been extended by Amaral *et al.* (1993). They show that, of the two different effects that have to be taken into account in interpreting the results, the second is modified by the impurity.
First there is the effect of spin disorder scattering; the resistance due to this is reduced by a magnetic field because the field tends to align the spins and so reduce the disorder.
Second there is the effect of weak localisation, which in their host alloy is dominated by spin–orbit scattering, so giving a positive magnetoresistance. The addition of magnetic impurities is then found to increase this positive contribution instead of reducing it, as one would expect if the only effect was to increase the dephasing of the partial waves by the scattering from magnetic impurities.
The reason for this apparent anomaly is that the localisation contribution to the magnetoresistance is due to both orbital and Zeeman splitting effects. The exchange interaction associated with the magnetic impurities enhances the paramagnetic susceptibility of the host alloy and thus, for a given field, increases the Zeeman splitting between the spin-up and the spin-down conduction bands (and hence their frequency difference). Thus the magnetic field 'kills' the enhanced resistance more rapidly than before and so increases the low-field magnetoresistance.
2. For further details and references to experimental work on metallic glasses, see the review by Howson and Gallagher (1988).

References

Abrahams, E., Anderson, P. W., Licciardello, D. C. and Ramakrishnan, T. V., (1979) *Phys. Rev. Lett.* **42** 673.
Aharonov, Y. and Bohm, D., (1959) *Phys. Rev.* **115** 485.
Altshuler, B. L., Aronov, A. G. and Spivak, B. Z., (1981) *Soviet Phys. JETP Lett.* **33** 94.
Altshuler, B. L., Aronov, A. G., Larkin, A. I. and Khmelnitzkii, D. E., (1981) *Soviet Phys. JETP* **34** 411.
Altshuler, B. L., Aronov, A. G., Spivak, B. Z., Sharvin, D. Yu. and Sharvin, Yu. V., (1982) *Soviet Phys. JETP Lett.* **35** 588.
Altshuler, B. L. and Aronov, A. G., (1985) in *Electron–Electron Interactions in Disordered Systems*, eds A. L. Efros and M. Pollak (North Holland).
Altshuler, B. L. and Lee, P. A., (1988) *Physics Today* **41** No. 12 (Dec) 36.
Amaral, V. S. *et al.*, *Europhys. Lett.* (1993) **21** 61.
Bergmann, G., (1983) *Phys. Rev. B* **28** 2914.
Bergmann, G., (1984) *Phys. Rep.* **107** 1.
Bergmann, G., (1987) *Phys. Rev. B* **35** 4205.
Bieri, J. B., Fert, A., Creuzet, G. and Schuhl, A., (1986) *J. Phys. F* **16** 2099.
Cahn, R. W., (1980) *Contemp. Phys.* **21** 43.
Cottrell, A., (1988) *Introduction to the Modern Theory of Metals* (Institute of Metals).
Drewery, J. S. and Friend, R. H., (1987) *J. Phys. F* **17** 1739.
Dugdale, J. S., (1977) *The Electrical Properties of Metals and Alloys* (Edward Arnold).
Duwez, P., (1981) in *Glassy Metals 1*, eds H-J. Güntherodt and H. Beck (Berlin, Springer-Verlag).
Elliot, S. R., (1984) *Physics of Amorphous Materials* (Longman).
Faber, T. E., (1969) in *The Physics of Metals*, ed. J. M. Ziman (Cambridge University Press).
Fukuyama, H., (1985) in *Electron–Electron Interactions in Disordered Systems*, eds A. L. Efros and M. Pollak (North Holland).
Fukuyama, H. and Hoshino, K., (1981) *J. Phys. Soc. Japan* **50**, 2131.
Gallagher, B. L., (1981) *J. Phys. F* **11** L207.
Gallagher, B. L., Greig, D. and Howson, M. A., (1984) *J. Phys. F* **14** L225.
Gallagher, B. L. and Hickey, B. J., (1985) *J. Phys. F* **15** 911.
Gey, W. and Weyhe, S., (1992) *Europhys. Lett.* **18** 331.

Greig, D., Gallagher, B. L., Howson, M. A., Law, D. S-L., Norman, D. and Quinn, F. M., (1988) *Mat. Sci. Eng.* **99** 265.

Güntherodt, H-J., Müller, M., Oberle, R., Hauser, E., Künzi, H. U., Liard, M. and Müller, R., (1978) *Institute of Physics Conference Series No. 39* (Bristol, London: Institute of Physics) p. 436.

Güntherodt, H-J. *et al.*, (1980) *J. Physique C8* **41** 381.

Hickey, B. J., Greig, D. and Howson, M. A., (1986) *J. Phys. F* **16** L13.

Hickey, B. J., Greig, D. and Howson, M. A., (1987) *Phys. Rev. B* **36** 3074.

Howson, M. A. and Gallagher, B. L., (1988) *Phys. Rep.* **170** 265.

Howson, M. A. and Greig, D., (1984) *Phys. Rev. B* **30** 4805.

Howson, M. A., Hickey, B. J. and Shearwood, C., (1986) *J. Phys. F* **16** L175.

Howson, M. A., Morgan, G. J., Paja, A. and Walker, M., (1988) *Z. Phys. Chem.* NF **157** 693.

Kaiser, A. B., (1982) *J. Phys. F* **12** L223.

Kaveh, M. and Wiser, N., (1984) *Advances in Physics* **33** 257.

Khmelnitzkii, D. E., (1984) *Physica B and C* **126** 235.

Kramer, B. and MacKinnon, A., (1993) *Reports on Progress in Physics* **56** 1469.

Larkin, A. I., (1980) *Sov. Phys. JETP Lett.* **31** 219.

Lee, P. A. and Ramakrishnan, T. V., (1985) *Rev. Mod. Phys.* **57** 287.

Lindqvist, P., (1992) *J. Phys. Condens. Matter* **4** 177.

Lindqvist, P. and Rapp, Ö., (1988) *J. Met. Phys.* **18**, 1979.

Lindqvist, P., Rapp, Ö., Sahnoune, A. and Ström-Olsen, J. O., (1990) *Phys. Rev. B* **41** 3841.

Maki, K., (1968) *Prog. Theor. Phys.* **39** 897.

Mayeya, F. M. and Howson, M. A., (1992) *J. Phys. Condens. Matter* **4** 9355.

Mitzutani, U. and Matsuda, T., (1983) *J. Phys. F* **13** 2115.

Mitzutani, U., Sasaura, M., Yamada, Y. and Matsuda, T., (1987) *J. Phys. F* **17** 667.

Movaghar, B. and Cochrane, R. W., (1991) *phys. stat. sol. (b)* **166** 311.

Mooij, J. H., (1973) *phys. stat. sol. A* **17** 321.

Mott, N. F. and Jones, H., (1936) *The Theory of the Properties of Metals and Alloys* (Oxford 1936, reprinted by Dover Publications, New York).

Mott, N. F. and Kaveh, K., (1985) *Advances in Physics* **34** 329.

Nguyen-Mahn, D., Mayou, D., Morgan, G. J. and Pasturel, A., (1987) *J. Phys. F* **17** 999.

Rainer, D. and Bergmann, G., (1985) *Phys. Rev. B* **32** 3522.

Rapp, Ö., (1993) *Europhys. News* **24** 102.

Rossiter, P. L., (1987) *The Electrical Resistivity of Metals and Alloys* (Cambridge University Press).

Sadoc, J. F. and Dixmier, J., (1976) *Mat. Sci. Eng.* **23** 187.

Sahnoune, A., (1992) Thesis, McGill University, Montreal.

Sahnoune, A., Ström-Olsen, J. O. and Fischer, H. E., (1992) *Phys. Rev. B* **46** 10035.

Schulte, A., Eckert, A., Fritsch, G. and Luscher, E., (1984) *J. Phys. F* **14** 1877.

Schulte, A. and Fritsch, G., (1986) *J. Phys. F* **16** L55.

Sharvin, D. Yu. and Sharvin, Yu. V., (1981) *Soviet Phys. JETP Lett.* **34** 272.

Thompson, R. S., (1970) *Phys. Rev. B* **1** 327.

Weir, G. F., Howson, M. A., Gallagher, B. L. and Morgan, G. J., (1983) *Phil. Mag. B* **47** 163.

Ziman, J. M., (1979) *Models of Disorder* (Cambridge University Press).

Index

Aharonov–Bohm effect, 111, 121
  with conduction electrons, 111–14
anomalous dispersion
  in electron wavefunctions, 90–4, 97
  in optics, 87–90
anti-localisation, 128

Boltzmann equation, 26, 33
Boltzmann theory of electrical conduction,
  25–32
Born approximation, 36, 39, 51

Ca–Al metallic glasses
  band structure, 212–13
  characteristic parameters, 218–19
  conductivity of, 211–23
  dephasing times, 222
  electronic heat capacity and Hall
    coefficient, 213
  magnetic impurities in, 222
  magnetoresistance, 213–23
  resistivity, temperature dependence,
    222–3
    effect of spin–orbit scattering on, 220–2
  value of F, 218–19, 221
closed paths, probability of, 115–16, 157–9
coherence of electron wavefunction, 111–12,
  115–20
collective electron modes, 70
comparison of experiment and theory,
  200–24
conductivity, electrical (see also resistivity),
  calculation by Boltzmann theory, 25–32
  of Ca–Al metallic glasses, 211–23
  of Cu–Ti metallic glasses, 201–5
  and diffusion, 32–3
  effect of Coulomb interaction on, 167–75
    Cooper channel, 174
    particle–hole channel, 167–70
  Einstein relation, 32

and weak localisation, 105–39
temperature dependence due to weak
  localisation, 117–18
conduction electrons, 21–3
Cooper channel interactions, 162–4, 174
  and magnetoresistance, 181–2
Coulomb interaction between conduction
  electrons, 143–62
Coulomb anomaly (see also enhanced
  interaction effect), 140–75
Coupling constant in exchange interaction,
  154–6
  for Hartree terms, 159–60, 230–1
  in Cooper channel, 162–4, 174,
    229–31
Cu–Ti metallic glasses, conductivity of,
  201–5

Debye–Waller factor, 58–60
density of states (for free electrons),
  24–5
  changes due to enhanced electron
    interaction, 141 et seq.
  changes with temperature, 165
dephasing processes, 117–18, 120–3, 207,
  211
  characteristic time in Ca–Al glasses, 222
diffusion, effect on electron wavefunctions,
  149–55
  importance of, in enhanced interaction
    effect, 161–2
diffusion coefficient of electrons, 32–3, 115
  et seq., 200–1
  values in Ca–Al glasses, 216, 218–20
  values in Cu–Ti glasses, 205
  values in a range of metallic glasses, 207
diffusive screening, 154–5, 171

Einstein relation, 32, 165, 200

238